Carvel S. Wolfe
U.S. Naval Academy

LINEAR PROGRAMMING with FORTRAN

Scott, Foresman and Company
Glenview, Illinois London

Library of Congress Catalog Card Number: 72-86897.

AMS 1970 Subject Classification 90C05.

ISBN: 0-673-07797-7.

Printed in the United States of America.
Regional offices of Scott, Foresman are located in
Dallas, Oakland, N.J., Palo Alto, and Tucker, Ga.

PREFACE

This book is intended as an introduction to the principal topics of linear programming. Students in the fields of operations analysis, computer science, analytical management, economics, game theory, government, and other related areas find that today most of their large problems are solved on electronic digital computers. Many of these problems use the elegant techniques of linear programming. The prime objective of this book is to make such techniques available to the student without subjecting him to unnecessary advanced mathematics.

The beauty of linear programming is found in its simple derivation. Only elementary mathematics is needed. Here, after the elementary derivations have been carried out, the results are stated in the form of algorithms. In turn, the algorithms lead naturally to computer routines that perform the necessary computations. One of the goals has been to develop a convenient notation that facilitates the computer programming and makes the answers readable at a glance.

All of this material has been used and refined for a number of years in the linear programming course at the U.S. Naval Academy. By the end of the course the students have been proud of their ability to run their own programs on high speed computers and to solve a variety of linear problems.

The linear algebra used in this book is explained as it is needed, and the only place that calculus appears is in Section 5.3 which may be omitted. An introduction to the local computer or "time sharing" system to be used is helpful. Over the past years the midshipmen have run the course problems on a number of I.B.M. machines at our computer center and also have used a half dozen of the area time sharing systems via remote teletypes. Currently we operate our own time sharing system.

The text includes worked out examples to illustrate each concept developed. The problem exercises are considered an integral part of the text. There are a number of short problems that can be done by hand while the student is learning technique. Then a considerable variety of problems is given to create interest and challenge ingenuity. Some of these exercises extend the material in the text and offer new ideas. The course can thus be adapted to many levels of preparation. The longer problems are expected to be run on a digital computer. To aid in the computer programming, FORTRAN subroutines that will carry out the necessary calculations are presented and explained in Chapter 4. Further help is found in the FORTRAN Glossary and the Appendices. It has been my practice to present Chapters 3 and 4 simultaneously so that as the student learns the Simplex Method he can discover the programming to carry out the algorithms.

The text is appropriately divided so that shorter courses may be developed. If only a half semester is available, Chapters 1 through 4 plus Chapter 10 offer the Simplex Method and the Transportation Algorithm. With more time the important ideas of duality in Chapters 5 through 7 may be added. Chapter 5 introduces game theory and solves matrix games. Chapters 8 and 9 may be considered separately or interchanged as desired. They can be taken up immediately after the Dual Simplex Method. A special effort has been made to simplify the ideas needed for integer programming and to improve the existing algorithms.

I am indebted to the many earlier authors in linear programming. My thanks to Peter Lax who read the manuscript and made valuable suggestions. Special thanks is due to my colleague, Harold Kaplan, who prepared the FORTRAN Glossary.

Carvel S. Wolfe

CONTENTS

1

GRAPHICAL METHODS

1.1 Introduction

Linear programming is increasingly relied upon in industry and government. The techniques of linear programming have found the cheapest, fastest, or most profitable way for large enterprises to function. They have also told government how to move supplies, allocate personnel, or use resources in the most efficient manner. In general, these techniques have helped to determine optimum choices in numerous competitive situations.

A part of the popularity of linear programming is due to our increased use of computers. The simultaneous development of fast computers has provided the necessary equipment to handle large problems that previously could not be solved because of lack of time or money. All that was necessary after the technical advance was the discovery and development of a theory that could be programmed on a digital computer. George B. Danzig proposed such a theory in 1947 while doing research for the U.S. Department of the Air Force. He originated the *Simplex Method* for solving linear problems. The first run of a linear program by Danzig's method occurred in 1952 on the computer of the National Bureau of Standards in Washington, D.C. Since then the method has been programmed for practically all large computers. Computer routines that may be adapted to almost any machine are found in Chapter 4.

In this chapter the terminology of linear programming will be explained and a graphical analysis used to solve two-dimensional cases.

1.2 The Language of Linear Programming

Throughout this book we will deal exclusively with linear problems called **linear programs**. As the word *linear* suggests, all the variables of a linear program are first degree. A *linear program* is characterized by a linear function of m variables that is to be maximized or minimized subject to a set of n linear constraints. A *constraint* is an equality or weak inequality[1] representing a restriction on the variables of a program.

Since the variables in a linear program usually stand for physical objects, they are most often required to be nonnegative. Thus a typical linear program is stated as: Find $x_i \geq 0$, $i = 1 \ldots, m$, that will maximize (minimize)

$$\langle \mathbf{1} \rangle \qquad \sum_{i=1}^{m} c_i x_i \qquad \text{such that}$$

$$\langle \mathbf{2} \rangle \qquad \sum_{i=1}^{m} a_{ij} x_i \leq b_j \qquad \text{for } j = 1, \ldots, n. \qquad \text{(Footnote 2)}$$

The x_i's are the variables, while a_{ij} and b_j in the n linear constraints of $\langle \mathbf{2} \rangle$, as well as the c_i in the linear function of $\langle \mathbf{1} \rangle$, are given constants for all subscripts. Geometrically the x_i's are the *coordinates of a point* (a vector) in m-dimensional space.

Definition 1. The coordinates of a point that satisfies all n linear constrains in $\langle \mathbf{2} \rangle$ constitute a **solution**.

Definition 2. If the coordinates of a solution point are all nonnegative, then the solution is called **feasible**.

Definition 3. If one or more of the coordinates of a solution point are negative then the solution is an **infeasible solution**. An **infeasible linear program** is a problem in which all solution points are infeasible.

Definition 4. The set of all feasible points is known as the **feasible region** of the problem.

Definition 5. If among the feasible solutions there is one for which the linear function in $\langle \mathbf{1} \rangle$ has its maximum (minimum) value, then this is called an **optimal feasible solution**.

[1] A *weak inequality* may possibly be satisfied by equality.

[2] $\sum_{i=1}^{m} x_i = x_1 + x_2 + \cdots + x_m$.

Definition 6. The linear function in $\langle \mathbf{1} \rangle$ is the **objective function** and its value at a point in the feasible region is referred to as the **objective value**.

From Definitions 5 and 6 we have that an optimal feasible solution is a feasible solution for which the objective value is maximum (minimum).

Later on we will generalize the above format to include constraints that are equations, and variables that are unrestricted in sign.

1.3 Graphical Analysis: Maximizing

Consider the following linear program.

Example 1. Fretmor is worried about his two toughest subjects, calculus and chemistry. After taking into account the rest of his subjects he realizes he must get at least 3 grade points out of calculus and chemistry combined to maintain his class standing. Of course, if he could pull an A in either one, he would get the maximum of 4 grade points. Experience has shown Fretmor that for each grade point in calculus he must study 10 minutes a day and for each grade point in chemistry he must study 15 minutes a day. Fretmor is unwilling to budget more than a total of 70 minutes a day for these two subjects. Under these conditions what is the maximum number of grade points he can earn and how should his time be divided between calculus and chemistry?

Solution. First choose the unknowns. Let x equal the number of grade points he receives in calculus and let y equal the number of grade points he receives in chemistry. The condition that he must get at least 3 grade points imposes the following constraint on the unknowns:

$$\langle \mathbf{1} \rangle \qquad x + y \geq 3.$$

This is a weak inequality and allows the sum of the grade points to be equal to 3 or greater than 3. The fact that 8 is the greatest number of grade points attainable imposes an upper bound constraint:

$$\langle \mathbf{2} \rangle \qquad x + y \leq 8.$$

Inequality $\langle \mathbf{2} \rangle$ will be called a **type I inequality** ($\leq$) while inequality $\langle \mathbf{1} \rangle$ will be called a **type II inequality** ($\geq$). Both type I and type II will allow the case of equality to be true.

Further constraints on the unknowns are

$$\langle \mathbf{3} \rangle \qquad x \leq 4, \qquad y \leq 4,$$

since 4 is the greatest number of grade points possible in one subject, and

$$\langle 4 \rangle \qquad x \geq 0, \qquad y \geq 0,$$

since zero is the least number of grade points possible. Constraints ⟨**4**⟩ are called nonnegativity constraints. A final constraint imposed by the time requirement is

$$\langle 5 \rangle \qquad 10x + 15y \leq 70.$$

This inequality may be simplified by dividing both sides by 5,

$$2x + 3y \leq 14.$$

The object of the problem is to maximize the number of grade points, so we find the largest possible number M such that

$$\langle 6 \rangle \qquad M = x + y.$$

The linear program is formulated as: Find x and y such that $M = x + y$ is as large as possible and such that

$$\langle 7 \rangle \qquad \begin{aligned} x + y &\geq 3 \\ x + y &\leq 8 \\ x &\leq 4 \\ y &\leq 4 \\ 2x + 3y &\leq 14, \end{aligned}$$

wherc x and y are nonnegative real numbers.

To solve the problem graphically we graph the region of the Cartesian plane corresponding to each of the constraints in ⟨7⟩ and ⟨**4**⟩. First draw the straight line $x + y = 3$, a line of slope -1 through (3, 0). A linear inequality in two variables is a half plane. The equality or straight line is called the **boundary** of the half plane. The half plane along with its boundary is called a **closed half plane**. We must decide on which side of the line $x + y = 3$ our half plane lies. An easy way is to solve the inequality for y:

$$y > 3 - x.$$

For fixed x the ordinates satisfying this inequality are larger than the corresponding ordinate on the line, and thus the inequality is satisfied for all points above the line. The region is shown in Figure 1–1.

Likewise graph the line $x + y = 8$ in the same Cartesian plane and shade the closed half plane $y \leq 8 - x$ which lies on or below the line. Since we must satisfy both constraints simultaneously we want the set intersection of these two closed half planes, or the strip between the two parallel lines as shown in Figure 1–2.

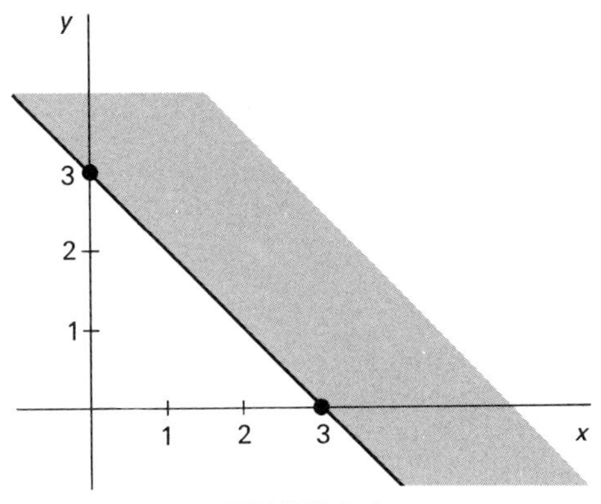

FIGURE 1–1

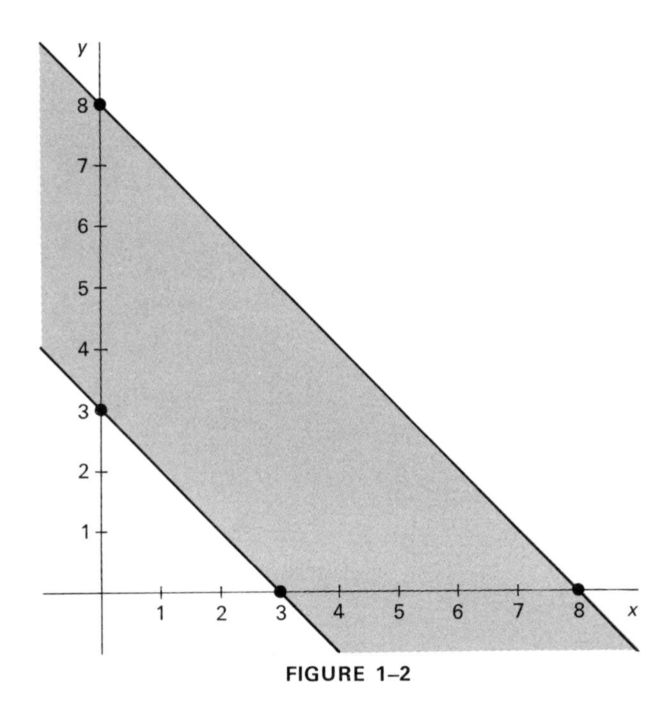

FIGURE 1–2

We are restricted to the first quadrant portion of this strip because x and y are nonnegative. Now intersect the first quadrant strip with the closed half planes $x \leq 4$ and $y \leq 4$ as shown in Figure 1–3.
Note from this graph that the constraint $x + y \leq 8$ is redundant. We get the same shaded region if $x + y \leq 8$ is omitted. This frequently happens in a linear program but seldom causes any trouble.

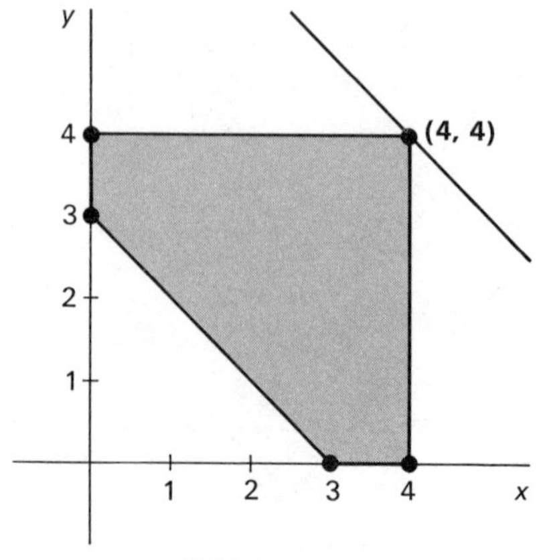

FIGURE 1–3

Finally intersect this region with the closed half plane $y \leq \dfrac{14 - 2x}{3}$ which is on or below the line $2x + 3y = 14$ of slope $-\frac{2}{3}$. The shaded region in Figure 1–4 represents the feasible region of the linear program.

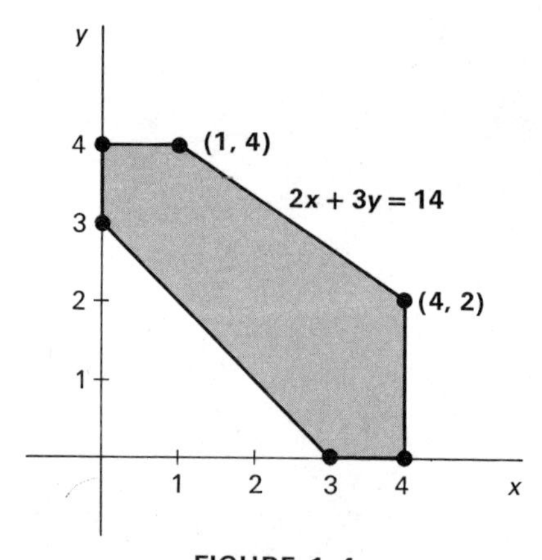

FIGURE 1–4

In our problem the objective function is $f(x, y) = x + y$. If we consider M as a parameter and graph $M = x + y$ for various values of M, we have a family of parallel lines with slope -1. Some of these lines will intersect the feasible region and contain many feasible solutions while others will miss and contain no feasible solution. We wish to find the line of this family that intersects the feasible region and is farthest out from the origin. The farther from the origin the greater will be the value of M.

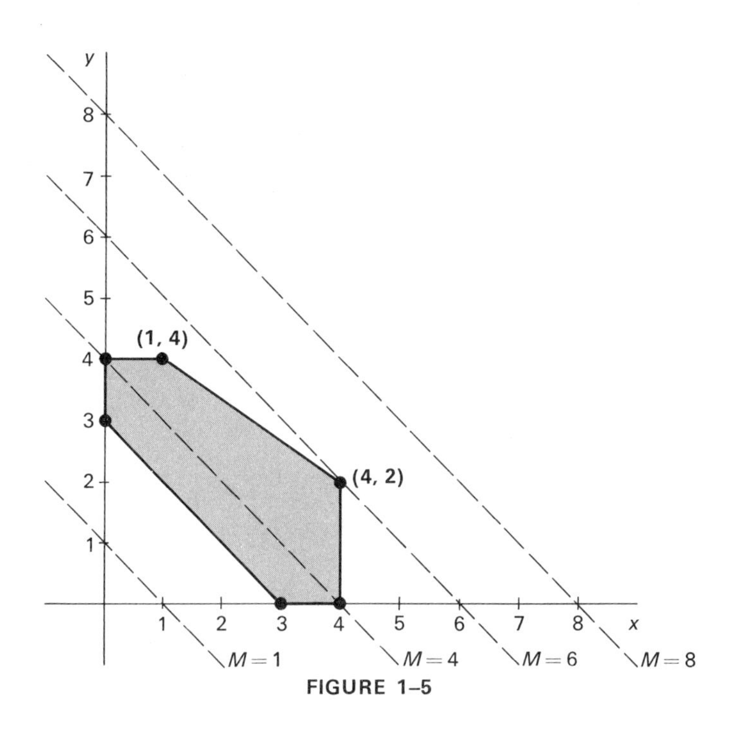

FIGURE 1–5

From the graph in Figure 1–5 we see that $M = 6$ does the job. It intersects the feasible region at the vertex (4, 2). Since (4, 2) is the only feasible point on $M = 6$, this solution is unique. Thus, Fretmor can achieve a maximum of 6 grade points, 4 for an *A* in calculus and 2 for a *C* in chemistry, by studying 40 minutes a day on calculus and 30 minutes a day on chemistry, using up his top allotted time of 70 minutes per day. ●[3]

1.4 Graphical Analysis: Minimizing

Consider another example of a linear program, a case in which the objective function is to be minimized.

Example 2. Candidate Wynnles is trying to hold a strict budget and decide how many personal appearances versus TV appearances he should make in the state presidential primary. His committee has figured that the total expense of each personal appearance at a campaign rally is \$15,000 while the cost of each TV speech is \$12,000. Mr. Wynnles has an efficiency expert who has estimated that each personal appearance

[3] ● indicates the end of an example.

rally will net 30,000 votes while each TV appearance will net 40,000 votes. Candidate Wynnles knows that he needs at least 240,000 votes to swing the state. Each of his personal appearance trips uses up 2 days while a live TV run takes only 1 day. He will spend at least 10 days in this state. For each personal appearance the party will provide him with 50 precinct workers but for a TV appearance the party can guarantee only 30 precinct workers. The local party boss claims that the winner must have at least 290 precinct workers. What is the minimum cost at which candidate Wynnles can carry this state?

Solution. Let x equal the number of personal appearances and y equal the number of TV appearances. The constraint on the number of votes to win is a type II inequality:

$$30{,}000x + 40{,}000y \geq 240{,}000.$$

Likewise we have a lower bound constraint on the number of days to be spent:

$$2x + y \geq 10.$$

The final constraint on the number of precinct workers is also of type II:

$$50x + 30y \geq 290.$$

The objective function is the total cost of the campaign:

$$\$15{,}000x + \$12{,}000y.$$

After the inequalities and objective function are simplified by dividing each through by a positive constant, the problem may be restated as follows: Find nonnegative numbers x and y such that

$$\begin{aligned} 3x + 4y &\geq 24 \\ 2x + y &\geq 10 \\ 5x + 3y &\geq 29, \end{aligned} \qquad \langle 1 \rangle$$

where $m = 5x + 4y$ is to be a minimum.

As in the first example we graph the closed half plane corresponding to each constraint and find the intersection of these half planes in the first quadrant. See Figure 1–6. Note in each case the half plane lies above its boundary line. The feasible region, as indicated by the shading in Figure 1–6, is infinite in extent. It includes all of the first quadrant above the boundary lines. A maximum in this case would be impossible to find, but we are looking for a minimum. Let us consider m a parameter in $m = 5x + 4y$ and draw the line for various values of m. We seek a line in this family of lines that intersects the feasible region and at the same time is as close as possible to the origin. Note in Figure 1–7 that the appropriate line is $m = 32$, and that the optimal

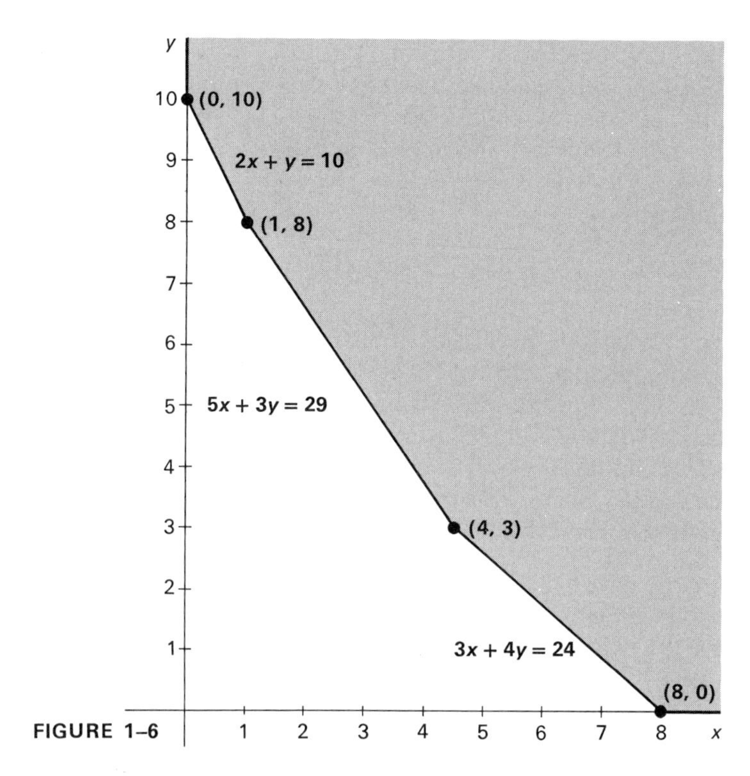

FIGURE 1–6

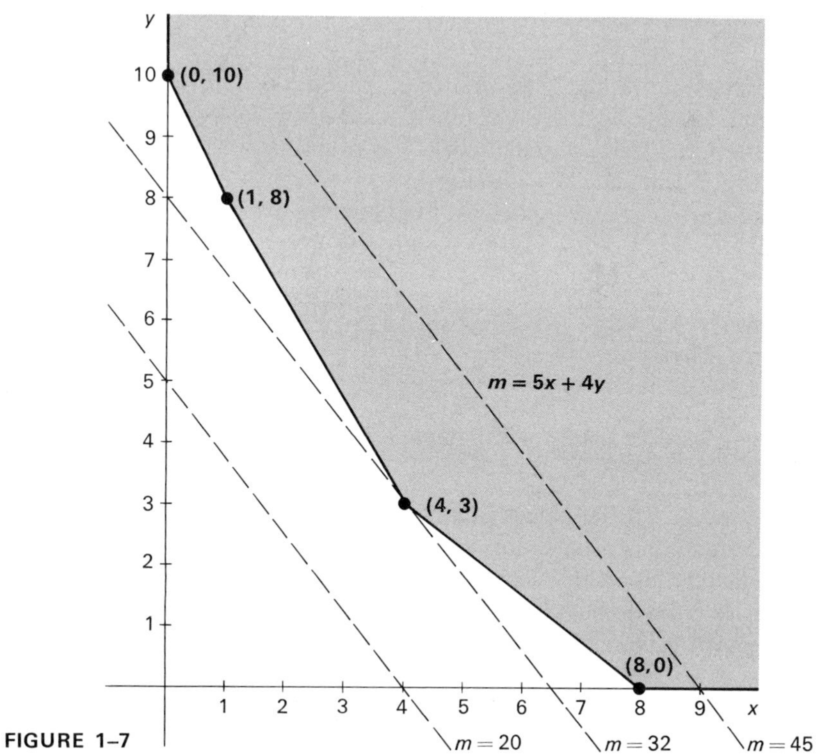

FIGURE 1–7

feasible solution occurs at vertex (4, 3). Each of the vertices was found by solving the corresponding pair of linear equations for their intersection. The optimal point (4, 3) tells us that candidate Wynnles should make 4 personal appearances and also 3 TV speeches to win the primary. By converting back to the original units we compute his minimum cost to be

$$\$15{,}000(4) + \$12{,}000(3) = \$96{,}000.$$

We may also check the original constraints to see that the optimal solution produces the required 290 precinct workers and the minimum of 240,000 votes. However, the remaining constraint shows that an additional day of Mr. Wynnles' time is required above the minimum 10 days. This additional day is known as the slack in that problem constraint. It is the quantity that must be added to or subtracted from an inequality to force it to equality. ●

Once again our solution is unique, but this is not necessarily the case as the next example will show.

1.5 Multiple Solutions

Example 3. Suppose in Example 2, page 7, the constraints remain the same but the objective function is changed by charging \$12,000 for a personal appearance and \$16,000 for TV time. Now what is the minimum cost for candidate Wynnles to win?
The problem may be stated as: Find nonnegative numbers x and y such that

$$\begin{aligned} 3x + 4y &\geq 24 \\ 2x + y &\geq 10 \\ 5x + 3y &\geq 29, \end{aligned}$$

where $m = 3x + 4y$ is a minimum.

Solution. The feasible region is the same as before but note in Figure 1–8 that one member of the family $m = 3x + 4y$ passes through a boundary line. This means that there are many solutions. Not only will vertex (4, 3) do but also the vertex (8, 0). Actually any point on the line segment between these two vertices is an optimal feasible solution.

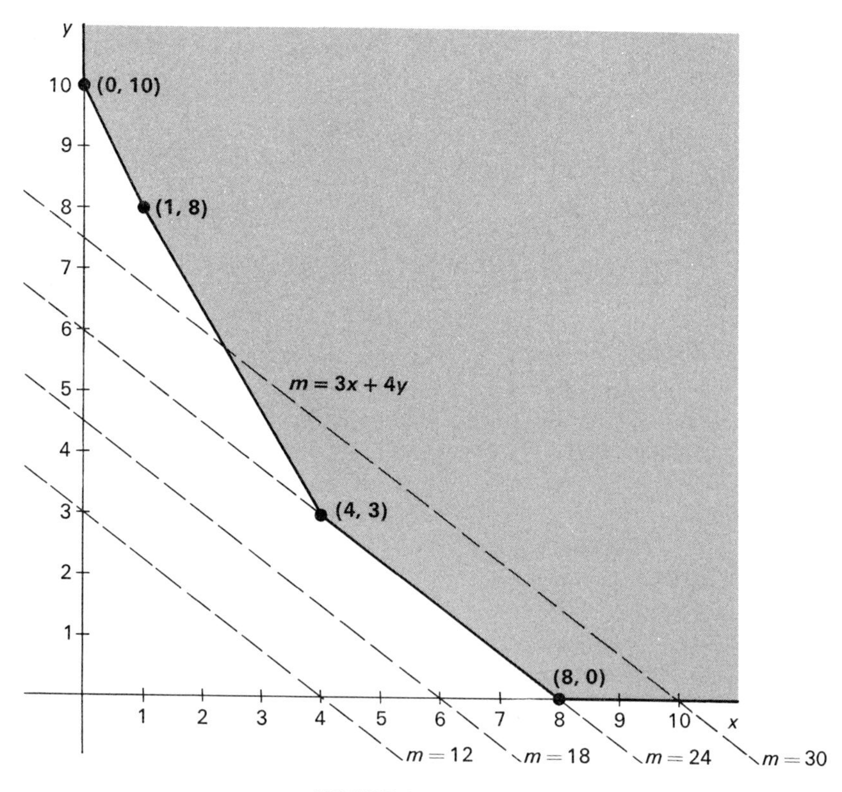

FIGURE 1–8

If we convert back to cost, the vertex (4, 3) corresponds to

$$\$12{,}000(4) + \$16{,}000(3) = \$96{,}000.$$

Vertex (8, 0) gives the same minimum cost,

$$\$12{,}000(8) + \$16{,}000(0) = \$96{,}000.$$

Any other point on this boundary gives the same minimum cost. For example $(6, \frac{3}{2})$ gives

$$\$12{,}000(6) + \$16{,}000(\tfrac{3}{2}) = \$96{,}000.$$

Of course from a practical viewpoint Mr. Wynnles can't make a fractional appearance so he would only be interested in integral answers. He could now choose the additional solution of 8 personal appearances and no TV appearances at the same cost. Both integral solutions produce the same number of votes but (8, 0) would require 16 days of his time. ●

Before proceeding to the theory, we may draw some inferences from our solutions in two dimensions. In order for the optimal solution to be *unique* it is necessary that the graph of the objective function touch the feasible region at a single vertex. The coordinates of this vertex make up the optimal feasible solution. It is possible for *many* optimal solutions to exist along a boundary of the feasible region. It is not possible to find an optimal solution strictly inside the feasible region. If the graph of the objective function passed through such an interior point, then the graph could be moved either in the direction of increasing M or decreasing m while still intersecting the feasible region. Finally if the feasible region is unbounded, it might be possible to move the objective graph arbitrarily far out from the origin while still passing through feasible points. In this case the corresponding objective value could be *unbounded.*

Problems Chapter 1

For the first three problems draw the graph of the region defined by the given linear constraints. For the sketched region find the maximum and the minimum of the objective function. Are all of the given constraints essential or may the number of constraints be reduced without changing the feasible region?

1–1. $x \geq 0,\ y \geq 0$
$2x + 3y \geq 11$
$3x + y \geq 6$
$4x - 5y \geq -30$
$5x + 2y \leq 45$
$4x - 3y \leq 13$
objective function $x + 3y$

1–2. $x \geq 0,\ y \geq 0$
$x + 2y \geq 4$
$x + 2y \leq 10$
$y - x \leq 5$
$x - y \leq 5$
objective function $2x + y$

1–3. $x \geq 0,\ y \geq 0$
$x + 2y \geq 4$
$x + 4y \leq 20$
$y \leq x + 5$
$x \leq y + 5$
objective function $x + y$

For the following problems use graphical methods to find the optimum solutions if such exist. Note whether or not the solutions are unique. In the case of multiple solutions give several answers.

1–4. $x \geq 0,\ y \geq 0$
$2x + y \geq 11$
$x + 2y \geq 10$
$3x + y \geq 13$
$x + 4y \geq 12$
Minimize $13x + 5y$.

1–5. Under the same constraints given in Problem 1–4, minimize $17x + 34y$. Does this function have a maximum over the feasible region?

1–6. $x \geq 0,\ y \geq 0$
$3x + 5y \leq 55$
$4x - 5y + 20 \geq 0$
$5x + 2y \leq 60$
Maximize $7x + 4y$; minimize $7x + 4y$.

1–7. Under the same constraints given in Problem 1–6, minimize $9x - 2y$. Does this function have a maximum? If so find it.

1–8. $x \geq 0,\ y \geq 0$
$3x - 7y \leq 9$
$3x - 7y \geq -14$
Maximize $6x - 21y$; is there a minimum value for this function over the feasible region?

1–9. $x \geq 0,\ y \geq 0$
$7x + 6y \geq 70$
$2x + 5y \leq 30$
$x - 2y \leq -10$
Maximize $ax + by$ for any real constants a and b; are there any feasible points?

1–10. $x \geq 0,\ y \geq 0$
$5x + 3y \geq 21$
$3x - 5y \geq -35$
$5x - 4y \leq 35$
$5x + 6y \leq 85$
$x + 2y \geq 7$
Maximize $15x + 18y$.

1–11. For the same constraints given in Problem 1–10, minimize $15x + 18y$.

1–12. For the constraints given in Problem 1–10, find both the maximum and the minimum of $3x + 2y$.

1–13. Mrs. Coffman was given the following advice. She should supplement her daily diet with at least 6000 USP units of vitamin A, at least 195 mg. of vitamin C, and at least 600 USP units of vitamin D. Mrs. Coffman finds that her local drug store carries blue vitamin pills at 5 cents each and red vitamin pills at 4 cents each. Upon reading the labels she sees that the blue pills contain 3000 USP units of A, 45 mg. of C and 75 USP units of D, while the red pills contain 1000 USP units of A, 50 mg. of C, and 200 USP units of D. What combination of vitamin pills should Mrs. Coffman buy to obtain the least cost? What is the least cost per day?

1–14. Seeall Manufacturing Company makes color TV sets. They produce a bargain set that sells for $100 profit and a deluxe set that sells for $150 profit. Along the set assembly line the bargain set requires 3 hours while the deluxe set takes 5 hours. The cabinet shop spends one hour on the cabinet for the bargain set and 3 hours on the one for the deluxe set. Both sets use 2 hours of time for testing and packing. On a particular production run the Seeall company has available 3900 man-hours on the assembly line, 2100 man-hours in the cabinet shop, and 2200 man-hours in the testing and packing department. How many sets of each type should they produce and what is their maximum profit?

1–15. Mr. Wise is head of the sporting goods department of the Eure Department Store. He decides to cut his inventory of spinning rods and reels by having a gigantic sale on 80 rods and 75 reels. Mr. Wise figures he will sell the $5 rods at a loss for a "come on" while selling the $12 spinning reels at a profit. He knows that most customers buying a rod will also pick up a new reel to match. The problem is to determine how much of a loss he can take on rods and how much of a mark up he needs on reels to sell the lot at a maximum profit. Mr. Wise knows from experience that to sell all the rods and reels he must attract at least 400 customers with his advertisement. For each dollar loss on a rod the ad will attract 400 customers, but each dollar mark up on reels will discourage 100 possible buyers. The Eure store manager insists that a reel must sell at a profit at least as big as the loss on a rod. The mark up on a reel can not exceed $6 and still be competitive. Finally Mr. Wise realizes that a rod and reel together should not sell for more than $19 to be sure that most go in sets. At what price does Mr. Wise advertise rods and what is his selling price on reels assuming that all are sold? What is the profit from this sale?

1–16. In Problem 1–15 suppose only 65 rods and 70 reels actually sell. Then what should have been the selling prices and the profit?

1–17. The Neely Nut Company sells mixtures of shelled nuts. Currently on hand the company has 10,000 pounds of shelled Arizona pecans and 9000 pounds of shelled English walnuts. Under the label of *Nut Delight* they sell a cardboard box containing one pound of shelled Arizona pecans and $1\frac{1}{2}$ pounds of shelled English walnuts. Another popular seller under their patented brand name of *Walcans*, is a plastic bag containing two pounds of shelled Arizona pecans and $1\frac{1}{2}$ pounds of shelled English walnuts. The packaging equipment which can only handle one brand at a time requires a half minute per box to fill, seal and label the *Nut Delight*, while only 12 seconds per bag is required on the *Walcans*. The management is facing an impending employee strike in $37\frac{1}{2}$ working hours. They wish to realize the maximum profit possible before the strike. If their profit is 60 cents on a box of *Nut Delight* and 50 cents on a bag of *Walcans*, then how should Neely Nut Company package its nuts?

1–18. If the strike is settled between the employees and management of Neely Nut Company, then how should the nuts be packaged in the previous problem to maximize profit?

1–19. The Steeping Tea Company is preparing two popular brands of tea, brand *A* and brand *B*. The quality of tea is determined by the position of the leaf on a twig of the tea bush. Orange Pekoe is made from the first leaf after the bud at the tip of the twig. Pekoe is made from the second leaf, and Souchong First is made from the third leaf. Steeping Tea has perfected its own blends. Brand *A* is a blend of 50% orange pekoe, 25% pekoe, and 25% souchong first. Brand *B* is a blend of 30% orange pekoe, 25% pekoe, and 45% souchong first. To meet back orders they will blend at least 550 pounds of orange pekoe at a cost of $2.00 per pound, at least 375 pounds of pekoe at a cost of $1.20 per pound, and at least 505 pounds of souchong first at a cost of $.80 per pound. How should Steeping Tea Company blend this tea into brands *A* and *B* so as to minimize the total cost of the two brands? What is the minimum total cost?

1–20. The White Glove Catering Service has a problem in planning for the men's bowling league banquet. It has been decided that the main dish will be a choice between a seafood platter and prime roast of beef. The league secretary reports that at least 300 members will attend the banquet. The president says that at least 50 more of his bowlers are beef eaters than seafood eaters. The banquet chairman has requested that at least 90 seafood platters be prepared. The number of platters of prime roast of beef that the manager of White Glove will prepare can be no greater than

130 more than the number of seafood platters, because then seafood would be left over and wasted. If the actual cost of preparing the seafood platter is $1.75 and the cost of the prime roast of beef platter is $1.90, then what combination of the two would produce the minimum cost to White Glove? If the cost of seafood went up so that a seafood platter cost $1.95, then what combination of dinners would yield the minimum cost to White Glove?

2

CONVEXITY

2.1 Definition and Properties of Convex Sets

One of the reasons that we could draw the conclusions in Section 1–5 is that the feasible regions of linear programs are *convex regions.*

Definition 1. A **region** or **set** R is **convex** if and only if for every two points of R the line segment connecting those points lies entirely in R.

The definition assumes that set R contains at least two points. The special cases where R equals either the empty set or a set of only one point are also taken to be convex regions. A theorem that follows easily from Definition 1 gives the following significant result.

Theorem 2–1. *The intersection of any number of convex regions is convex.*

Proof. Start with a given collection of convex regions and let R be their intersection. R consists of all points common to each region in the collection. Thus if P_1 and P_2 are points belonging to R, they belong to each region of the original collection. Since each given region is convex, the segment P_1P_2 belongs to each given region and therefore P_1P_2 belongs to R. Now R satisfies Definition 1. ■[1]

In order to make Definition 1 workable, we need an algebraic expression for all the points on a line segment. This is most easily done by introducing a parameter, α, and writing the equations of the line segment in parametric form. Let X and Y be two arbitrary points in m-dimensional space with their coordinates given by x_i, $i = 1$ to m

[1] ■ indicates the end of a proof.

and y_i, $i = 1$ to m, respectively. Then the line segment XY is the set of all points U with coordinates satisfying the parametric equations

$$\langle 1 \rangle \qquad u_i = (1 - \alpha)x_i + \alpha y_i, \qquad i = 1, \ldots, m \qquad \text{where } 0 \leq \alpha \leq 1.$$

Any of the points U is called a *convex combination* of points X and Y. The set of all convex combinations of X and Y for $0 \leq \alpha \leq 1$ is precisely the line segment XY which is convex.

For the remainder of this chapter we will concentrate on the two-dimensional plane where the relationships are easy to see and draw. The ideas have to be extended to higher dimensional space in order to apply to general linear programs. In two dimensions if $X(x_1, x_2)$ and $Y(y_1, y_2)$ are points in a plane, then from the parametric equations $\langle 1 \rangle$ the line segment XY is the set of points $U(u_1, u_2)$ satisfying the pair of parametric equations

$$\langle 2 \rangle \qquad \begin{aligned} u_1 &= (1 - \alpha)x_1 + \alpha y_1 \\ u_2 &= (1 - \alpha)x_2 + \alpha y_2, \qquad \text{for } 0 \leq \alpha \leq 1. \end{aligned}$$

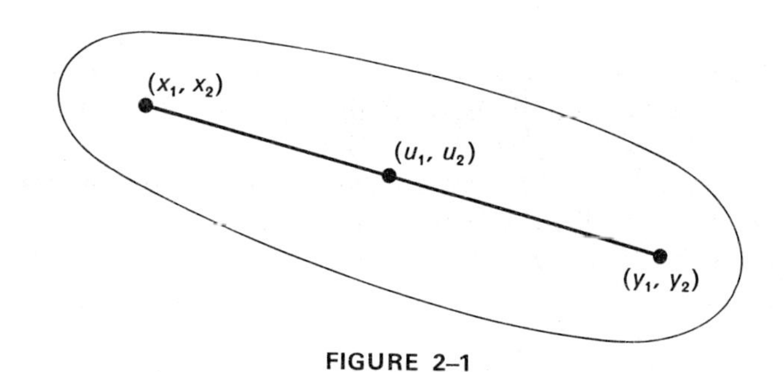

FIGURE 2–1

Let R be a plane convex region. The definition of convexity states that the set of points satisfying parametric equations $\langle 2 \rangle$ belongs to R whenever (x_1, x_2) and (y_1, y_2) belong to R. Note that the values 0 and 1 of the parameter α produce the end points X and Y respectively. The remaining real values of α between 0 and 1 produce points along XY at those proportionate distances from X. For example $\alpha = \frac{1}{2}$ results in the familiar midpoint formulas, while $\alpha = \frac{1}{3}$ will place U one-third of the way from X to Y. The line segment defined by equations $\langle 2 \rangle$ is called a *closed segment* because it includes its end points X and Y. An *open segment* would omit the end points. If α is an unrestricted real number, then equations $\langle 2 \rangle$ would represent the entire line defined by points X and Y.

We saw in Chapter 1 that a linear constraint $ax_1 + bx_2 \leq c$ defined a closed half plane where the boundary line corresponds to equality in the constraint.

Theorem 2–2. *The closed half plane $ax_1 + bx_2 \leq c$ is a convex region.*

Proof. Let $Y(y_1, y_2)$ and $Z(z_1, z_2)$ belong to the closed half plane $ax_1 + bx_2 \leq c$. Thus their coordinates satisfy

$$ay_1 + by_2 \leq c$$
$$az_1 + bz_2 \leq c.$$

Let $U\ (u_1, u_2)$ be any point on the line segment YZ. From the parametric equations of YZ we have

$$\begin{aligned} au_1 + bu_2 &= a[(1-\alpha)y_1 + \alpha z_1] + b[(1-\alpha)y_2 + \alpha z_2] \\ &= (1-\alpha)(ay_1 + by_2) + \alpha(az_1 + bz_2) \\ &\leq (1-\alpha)c + \alpha c, \qquad \text{since both } \alpha \text{ and } 1-\alpha \text{ are nonnegative} \\ &\leq c. \end{aligned}$$

This proves that all points on the line segment YZ are in $ax_1 + bx_2 \leq c$, and therefore the closed half plane is convex. ■

An open half plane is also convex, but is not of interest in linear programming. In m-dimensional space, $m > 2$, the correspondent to a closed half plane is a closed half space. The closed half space may be proved to be a convex region by allowing the subscripts in Theorem 2–2 to run from one to m.

Combining Theorems 2–1 and 2–2 gives the following theorem.

Theorem 2–3. *The intersection of any number of half planes is a convex region.*

Definition 2. A **polygonal convex region** is the intersection of a positive finite number of closed half planes.

The feasible region of a finite set of linear inequalities in two variables is the intersection of a finite number of closed half planes and is therefore a polygonal convex region.The feasible regions discussed in Chapter 1 were all polygonal convex regions.

Definition 3. A **convex polygon** is a polygonal convex region R such that any line through a point of R intersects R in a closed segment.

Since a closed segment has finite length, a convex polygon is a bounded region of the plane. The boundaries of a polygonal convex region are made up of either all or portions of the boundary lines of the intersecting planes. These boundaries are then either lines, rays or closed segments. To see that each case may occur refer to Figure 2–2.

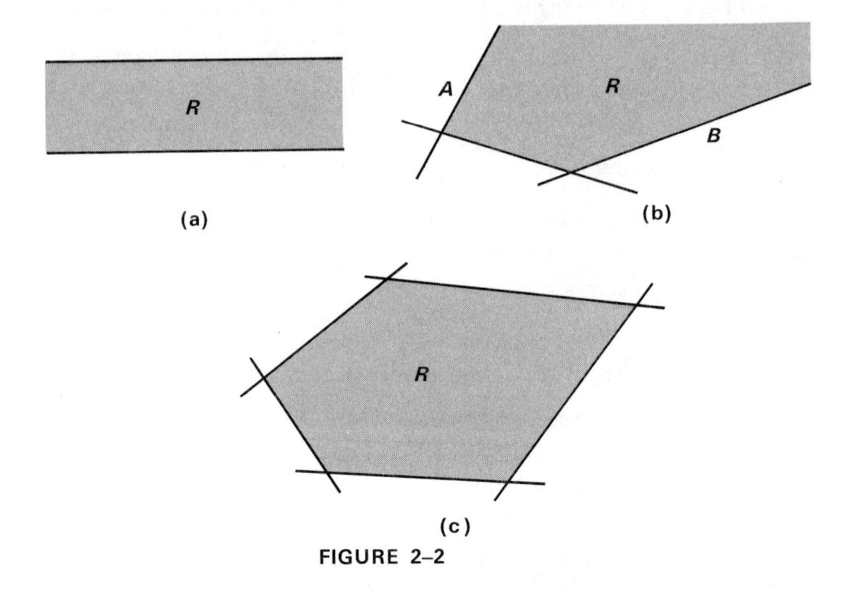

FIGURE 2–2

In Figure 2–2(a) the boundaries of R are parallel lines. In (b) boundaries A and B are rays or half lines. All of the boundaries in (c) are line segments. Note that (a) and (b) typify general polygonal convex regions that are unbounded or infinite in extent, and (c) illustrates a convex polygon that is necessarily a finite region of the plane.

The analog of a convex polygon in three dimensions is called a **convex polyhedron.** The faces of a convex polyhedron are convex polygons. All of the regular polyhedrons such as the tetrahedron and cube are convex. A characteristic of these figures is that they have corners, commonly called vertices or extreme points.

Definition 4. A **vertex** or extreme point in a convex set R is a point of R that cannot be written in equations $\langle \mathbf{1} \rangle$ as a convex combination of two other distinct points of R.

According to Definition 4 vertices in a sense stand alone. They cannot be reached by any line segment connecting two points of R that are distinct from the vertex itself. In the case of convex polyhedra, the vertices will be where edges come together. An interesting convex region is the solid sphere because every point of its surface is an extreme point.

In two dimensions a vertex of a polygonal convex region is a point in the region where two boundary lines join. It is possible for more than two boundary lines to pass through a vertex, but then one of them will be redundant in defining the convex region. We might just as well say that for a polygonal convex region a vertex is a point in the region on exactly two boundary lines.

2.2 Linear Functions

In the plane with coordinates x_1 and x_2 let us examine linear functions of the form $f(x_1, x_2) = ax_1 + bx_2$ where a and b are real constants not both zero. If $Y(y_1, y_2)$ and $Z(z_1, z_2)$ are distinct points of the plane, then the *parametric equations* of the line through Y and Z are

$$x_1 = (1-\alpha)y_1 + \alpha z_1$$
$$x_2 = (1-\alpha)y_2 + \alpha z_2, \qquad \text{for all real numbers } \alpha.$$

Let $f(y_1, y_2) = m$ and $f(z_1, z_2) = M$. For any point $X(x_1, x_2)$ on line YZ we have

$$\begin{aligned} f(x_1, x_2) &= a[(1-\alpha)y_1 + \alpha z_1] + b[(1-\alpha)y_2 + \alpha z_2] \\ &= (1-\alpha)(ay_1 + by_2) + \alpha(az_1 + bz_2) \\ \langle 3 \rangle \qquad &= (1-\alpha)f(y_1, y_2) + \alpha f(z_1, z_2) \\ &= (1-\alpha)m + \alpha M \\ &= m + \alpha(M - m). \end{aligned}$$

Using this result it is easy to prove the following theorem.

Theorem 2–4. *If f is linear and has the same value at two distinct points Y and Z, then f remains constant along the line through Y and Z. On the other hand if f has different values at Y and Z, then at each point of the open segment YZ, f will have a value strictly between its values at Y and Z.*

Proof. In equation $\langle 3 \rangle$ let $m = M$ and then $f(x_1, x_2) = m$ is constant. If $m \neq M$, say $m < M$, then $M - m > 0$. Equation $\langle 3 \rangle$ may be considered a function of parameter α. Call it $f(\alpha)$.

$$f(\alpha) = m + \alpha(M - m),$$
$$f(0) = m \qquad \text{and} \qquad f(1) = M.$$

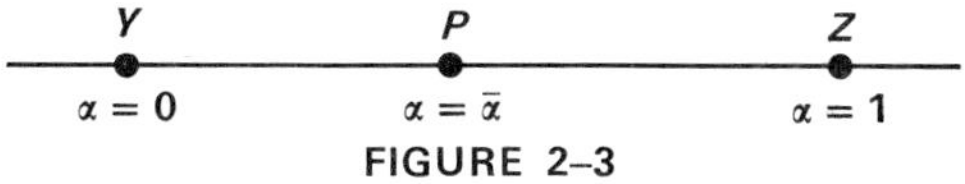

Y P Z
$\alpha = 0$ $\alpha = \bar{\alpha}$ $\alpha = 1$

FIGURE 2–3

For any two values of the parameter $\alpha_1 < \alpha_2$,

$$(M - m)\alpha_1 < (M - m)\alpha_2$$
$$m + (M - m)\alpha_1 < m + (M - m)\alpha_2$$
$$f(\alpha_1) < f(\alpha_2).$$

This means that f is a strictly increasing function of parameter α. That is, as α increases the functional values $f(\alpha)$ steadily increase. Now, let $P(x_1, x_2)$ with parameter value $\bar{\alpha}$ be any point of the open segment YZ as in Figure 2–3. In the parametric equations of segment YZ parameter α is also strictly increasing so $0 < \bar{\alpha} < 1$. Then,

$$f(0) < f(\bar{\alpha}) < f(1) \qquad \text{or} \qquad m < f(x_1, x_2) < M. \quad ■$$

A conclusion of Theorem 2–4 is that a linear function defined over a closed segment will assume its maximum and minimum at the end points of that segment.

2.3 The Extreme Point Theorem

We are particularly interested in the extreme points or vertices of a polygonal convex region. Fundamental in this development is the idea observed in Chapter 1 that a linear objective function attains its optimum over a convex region at a vertex of that region.

Theorem 2–5. *The maximum and minimum values of a linear function f defined over a convex polygon R exist*[2] *and are found at the vertices of R.*

Proof. Suppose $P(\bar{x}, \bar{y})$ is an interior point of R, that is a point of R not on the boundary of R. Let L be an arbitrary line through P which by Definition 3 intersects R in a closed segment. Let $P_1(x_1, y_1)$ and $P_2(x_2, y_2)$ be the end points of that segment. Since the remainder of line L is exterior to R, the end points P_1 and P_2 must be on the boundary of R. If the linear constraint defining the boundary through P_2 is $cx + dy \le e$, then $cx_2 + dy_2 = e$, while $c\bar{x} + d\bar{y} = k < e$, since P is an interior point. The linear function $cx + dy$ will assume values between k and e for points on L between P and P_2 by Theorem 2–4. Similar strict inequalities will be satisfied by all the points between P and any other boundary line. Thus all points between an interior point and a boundary point are interior points. This means the open segment

[2] The existence is also guaranteed by a theorem from calculus. A linear function is continuous and any continuous function defined over a closed bounded region attains its extreme values at points of the region.

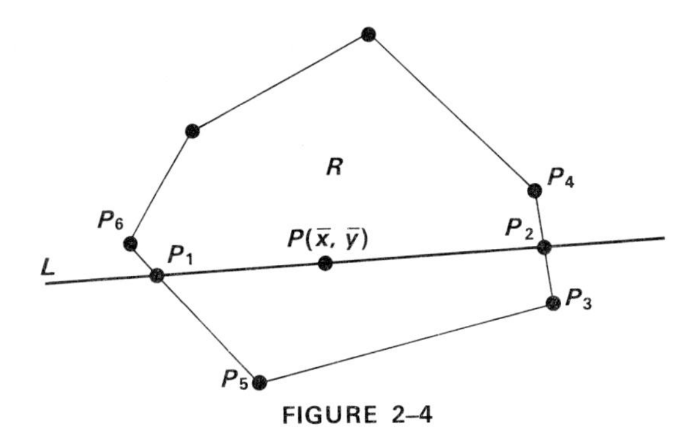

FIGURE 2–4

between P_1 and P_2 consists entirely of interior points and therefore L may intersect the boundary in only two points.

Let the given linear function be f. If $f(x_1, y_1) \leq f(x_2, y_2)$ then by Theorem 2–4

$$\langle \mathbf{1} \rangle \qquad f(x_1, y_1) \leq f(\bar{x}, \bar{y}) \leq f(x_2, y_2).$$

The boundaries of R are also closed segments by Definition 3. The end points of boundary segments lie on exactly two boundary lines, so they are vertices. Thus if P_2 is not a vertex, it will lie between two vertices, say P_3 and P_4. Applying Theorem 2–4 again, $f(x_2, y_2)$ is bounded on that boundary segment, say

$$\langle \mathbf{2} \rangle \qquad f(x_3, y_3) \leq f(x_2, y_2) \leq f(x_4, y_4).$$

Combining inequalities $\langle \mathbf{1} \rangle$ and $\langle \mathbf{2} \rangle$ gives

$$\langle \mathbf{3} \rangle \qquad f(\bar{x}, \bar{y}) \leq f(x_4, y_4).$$

By the same argument $f(x_1, y_1)$ is bounded by the values of f at two vertices, say

$$\langle \mathbf{4} \rangle \qquad f(x_5, y_5) \leq f(x_1, y_1) \leq f(x_6, y_6).$$

Finally combine inequalities $\langle \mathbf{1} \rangle$, $\langle \mathbf{3} \rangle$, and $\langle \mathbf{4} \rangle$ giving the result

$$\langle \mathbf{5} \rangle \qquad f(x_5, y_5) \leq f(\bar{x}, \bar{y}) \leq f(x_4, y_4).$$

We have shown that f at any interior point P is bounded by the values of f on the boundary and those in turn are bounded by values of f at vertices. Since there are only a finite number of vertices in a convex polygon we may examine f at each vertex and choose one with as large a value of f as possible and one with as small a value of f as possible. These choices of vertices may not be unique, but they do represent maximum and minimum points of f over R, completing the proof. ■

In the case of a general polygonal convex region that may be unbounded, a linear function over that region may increase or decrease without bound. The intersection of a line through a point of R with R may be a ray or the entire line. Along a ray at least one coordinate must continue to increase or decrease and thus any linear f involving that coordinate may either increase or decrease without bound. To state a theorem in this case, it is necessary to hypothesize the existence of an optimum. We state the theorem without proof as the reasoning follows the same arguments given in Theorem 2–5.

Theorem 2–6. *If the maximum or the minimum value of a linear function defined over a polygonal convex region exists, then it is found on the boundary or at a vertex of the region.*

If a boundary were an entire line, then there would be no vertex on that boundary. However, for a feasible region in the first quadrant this could not happen, so every boundary line contains at least one vertex. As long as a vertex is available on the optimum boundary the optimum value occurs at a vertex. Thus, the result is the same as in the case of a convex polygon provided the optimum exists.

The proofs in this chapter have been carried out in two dimensions for simplicity and ease of understanding. The ideas generalize to n dimensions and hold for the general linear programming problem.

Problems Chapter 2

2–1. Find a set of parametric equations for the line segment from $P_1(2, -5)$ to $P_2(-3, 4)$. What are the parametric equations of the entire line through P_1 and P_2? For what values of parameter α do you get points on the line below P_1, and for what values of α do you get points on the line above P_2?

2–2. On line segment $P_1(-2, -4)$ to $P_2(7, 2)$ find the coordinates of the point P that is $\frac{2}{3}$ of the way from P_1 to P_2. Check your result by using the distance formula from analytic geometry.

2–3. Prove that an open half plane is a convex set by following the proof to Theorem 2–2.

2–4. For the linear function $f = 2x_1 + 3x_2$ defined over segment $P(-5, 3)$ to $Q(6, -1)$ find the maximum value of f and the minimum value of f. Find an arbitrary point interior to the segment PQ. Show that the value of f at this point is between the values of f at P and Q. Check your result by computing f as a function of α and testing it for your value of the parameter.

2–5. If $f = -3$ at $Y(-4, -1)$ and $f = 5$ at $Z(2, 3)$ find in terms of the parameter α the linear function $f(\alpha)$ defined over line segment YZ. Let (x_1, x_2) be an arbitrary point on segment YZ and find f as a function of x_1 and x_2. Note f is linear and may involve only one of the variables, either x_1 or x_2. By eliminating the parameter find the rectangular equation of the line through X and Y.

2–6. If $f(\alpha) = 3\alpha + 5$, $0 \leq \alpha \leq 1$, is an increasing function defined over the line segment from $Y(-5, 1)$ to $Z(2, -6)$, find the value of f at the end points of the segment. Find the coordinates of the point (x_1, x_2) for which $f = \frac{15}{2}$.

2–7. Write out the details to the proof of Theorem 2–6 by following the outline of the proof in Theorem 2–5.

2–8. Prove that a linear function defined over a convex polyhedron has both a maximum and a minimum value that occur at vertices of the polyhedron. Hint: Consider a plane section through any interior point and apply Theorem 2–5 to the resulting convex polygon.

3

THE SIMPLEX METHOD

3.1 Gauss-Jordan Elimination

The word **simplex** comes from the concept of the simplest possible convex figure that has exactly one more vertex than the dimension of the space. The zero-dimensional simplex is a single point, while a line segment, with two end points, is a simplex in one dimension. In two dimensions with three vertices, a simplex is a triangle, and in three dimensions with four vertices, a simplex is a pyramid or tetrahedron. Each of these figures is the smallest convex set containing its vertices, and is known as the convex hull of the vertices. By definition the **convex hull** of any set of points is the smallest convex set containing those points. Whenever the number of vertices or extreme points is finite, then the convex hull of those points is called a **convex polyhedron**. The simplices are then special cases of convex polyhedra having one more vertex than the space dimension, and with the additional property that their faces or boundaries are simplices of the next lower dimension. Linear programming has lifted the word simplex from geometry and applied it to the process of locating an optimal vertex among the vertices of a feasible region.

In developing the Simplex Method, George Dantzig made use of the classical Gauss-Jordan elimination process. This elimination is familiar to anyone who has solved a system of linear equations. The key idea is to take a multiple of one equation and add it to or subtract it from another equation in order to eliminate one of the unknowns from the second equation. By removing unknowns, we hope to reduce the original system to an equivalent system for which it is easier to solve for the remaining unknowns. As an example, let us carry out this elimination for the following system of two equations in three unknowns in order to solve for x_1 and x_2 in terms of x_3.

Example 1.

$$2x_1 - 4x_2 + x_3 = 2$$
$$x_1 - 4x_2 - x_3 = 5$$

A notational convenience that is standard in linear programming is to place the coefficients of these equations into a rectangular block called a **tableau**. In linear algebra the rectangular array of numbers is called a **matrix**.

x_1	x_2	x_3	
(2)	−4	1	2
1	−4	−1	5

To solve this system for unknowns x_1 and x_2, we will eliminate x_1 from the second equation and x_2 from the first equation. Then the first equation is immediately solvable for x_1 and the second equation is solvable for x_2. Each of these eliminations is called a **pivoting iteration** or briefly a **pivot** of the tableau. The initial elimination amounts to taking a multiple, in our case $\frac{1}{2}$, of the first row and subtracting it from the second row. The operation revolves about the coefficient of x_1 in the first row. Thus this coefficient is called a **pivot** during the pivoting iteration. The pivots will be circled in our tableaux. The next tableau shows the first step in pivoting which is to divide the row in which the pivot occurs (**pivotal row**) by the pivot.

x_1	x_2	x_3	
1	−2	$\frac{1}{2}$	1
1	−4	−1	5

The pivot position is now 1 in the above tableau. The final step in pivoting is to produce zeros in all remaining positions of the column in which the pivot occurs (**pivotal column**). In our case, subtract the first row from the second row.

x_1	x_2	x_3	
1	−2	$\frac{1}{2}$	1
0	(−2)	$-\frac{3}{2}$	4

The result of pivoting is a unit vector in the pivotal column.

The second pivoting iteration is to eliminate the coefficient of x_2

from the first row. Since x_2 is in the second column, the new pivot must be in the second column, but not in the first row, so choose the second row. If we were solving for x_3 instead of x_2, then the second pivot would be in the third column of the second row. Divide the second row by the pivot giving the following tableau.

x_1	x_2	x_3	
1	-2	$\frac{1}{2}$	1
0	1	$\frac{3}{4}$	-2

Then produce zeros in the remaining positions of the second column by taking twice the second row and adding it to the first row.

x_1	x_2	x_3	
1	0	2	-3
0	1	$\frac{3}{4}$	-2

This leaves a unit vector in the x_2 column.

When the number of unit column vectors is equal to the number of equations the corresponding square array is called the **identity matrix.** By definition the identity matrix has ones down its main diagonal and zeros elsewhere.

Notice that each step in the pivoting process resulted in a system equivalent to the original system of equations in the sense that each possesses the same solution set. A solution for x_1 and x_2 may now be read from the final tableau.

$$\begin{aligned} 1x_1 + 0x_2 + 2x_3 &= -3 \\ 0x_1 + 1x_2 + \tfrac{3}{4}x_3 &= -2 \end{aligned} \qquad \text{or} \qquad \begin{aligned} x_1 &= -3 - 2x_3 \\ x_2 &= -2 - \tfrac{3}{4}x_3 . \end{aligned} \quad \bullet$$

Of the many solutions, the most important for our purposes is the one in which $x_3 = 0$. This will be called a *basic solution* and is easily read off from the final tableau. The values of the *basic variables* are found in the right-hand column opposite the ones of the unit vectors. Thus $x_1 = -3$ and $x_2 = -2$ is a basic solution.

In the general case consider m equations in n unknowns where $m \leq n$. We will set $n - m$ of the variables equal to zero and try to solve for the remaining m variables.

Definition 1. For $m \leq n$ the $n - m$ variables that are chosen to be zero are called **nonbasic** and the remaining m variables are called **basic.** A solution for the basic variables is called a **basic solution.** A **nonbasic**

solution is a solution to the problem that involves more than m nonzero variables.

Usually there are many basic solutions. In our example there are three basic solutions in which the basic variables are respectively (x_1, x_2), (x_1, x_3), and (x_2, x_3). The number of possible basic solutions may be found from the formula for the binomial coefficient

$$\binom{n}{m} = \frac{n!}{m!(n-m)!}.$$

A particular basic solution will exist and be unique provided the matrix formed from the coefficients of those basic variables is nonsingular. This square matrix of coefficients will be nonsingular if its determinant is nonzero.

3.2 The Extended Tableau

The tableaux in this section are called **extended tableaux** since they include an identity matrix. To illustrate the Simplex Method by use of the extended tableau consider the following linear program.

Example 2. Find nonnegative numbers x_1 and x_2 subject to the constraints

$$\begin{aligned} x_1 + 3x_2 &\le 7 \\ 2x_1 + x_2 &\le 4, \end{aligned}$$

such that the linear function $x_1 + 2x_2$ is a maximum. In order to convert the inequalities into equations we introduce nonnegative *slack variables* x_3 and x_4. The slack variables take up the slack in the constraint. That is, they add into the left side whatever is necessary to bring it up to equality. The maximum will not be changed if the slack variables are given zero coefficients in the objective function. Write the objective as $x_1 + 2x_2 + 0x_3 + 0x_4 = M$.

The problem may now be stated as follows. Find x_1, x_2, x_3, x_4 satisfying

$$\langle 1 \rangle \qquad x_1 \ge 0, \quad x_2 \ge 0, \quad x_3 \ge 0, \quad x_4 \ge 0,$$

such that

$$\langle 2 \rangle \qquad \begin{aligned} x_1 + 3x_2 + x_3 + 0x_4 &= 7 \\ 2x_1 + x_2 + 0x_3 + x_4 &= 4, \end{aligned}$$

and

$$\langle 3 \rangle \qquad M = x_1 + 2x_2 + 0x_3 + 0x_4$$

is a maximum.

We are looking for an optimal feasible solution. Fortunately, we need only look among the basic solutions of system ⟨**2**⟩ to find our answer. The basic solutions of system ⟨**2**⟩ are the solutions for any two of the variables while the remaining two variables are set equal to zero. The two constraining equations in ⟨**2**⟩, plus the additional two equations setting the nonbasic variables equal to zero, constitute four faces or hyperplanes in the four-dimensional solution space of x_1, x_2, x_3, and x_4. The intersection of these four hyperplanes is a vertex or extreme point in the solution space. In general, the intersection of at least n hyperplanes in n-dimensional space is a vertex. As we saw in Chapter 2 by the Extreme Point Theorem, a linear function attains its maximum at an extreme point of the feasible region. Since these extreme points correspond to basic solutions, we will find our maximum among the basic solutions. The result is stated in the following theorem.

Theorem 3-1. *If an optimal feasible solution exists, then at least one basic optimal feasible solution exists as well.*[1]

In the usual case, where the optimum is unique, Theorem 3–1 guarantees that it will be basic. In case the optimum is not unique we have what are called alternate optima. They need not be basic but at least one basic solution will have the same optimum value. As a matter of fact, unless the feasible region is unbounded, every nonbasic optimal feasible solution is a convex combination of two basic optimal feasible solutions.[2]

Returning to our example, the initial extended tableau is

	x_1	x_2	x_3	x_4	
	1	(3)	1	0	7
	2	1	0	1	4
M	−1	−2	0	0	0

[1] For a proof of this theorem by linear algebra see S. Vajda, *Mathematical Programming*, 1961, p. 12.
[2] See Vajda, ibid., p. 12.

The coefficients of the objective function have been transferred to the left side of equation ⟨**3**⟩ and placed in a row at the bottom of our tableau. The box in the lower right corner will contain the objective value for the basic variables provided those basic variables have zero coefficients in the bottom row. Choosing x_1 and x_2 to be nonbasic, $x_1 = x_2 = 0$, the initial basic feasible solution is $x_3 = 7$, $x_4 = 4$ with initial objective value $M = 0$. We may increase M by increasing any variable with a negative coefficient in the last row. The greatest increase in M per unit increase in x occurs by choosing the variable whose negative coefficient in the last row has the largest absolute value. This most negative entry is -2 and it determines the second column to be a pivotal column. We now try to find a pivotal row by asking the question, "How much may x_2 be increased?" Keep x_1 nonbasic, that is equal to zero, and consider the equations

$$x_3 = 7 - 3x_2$$
$$x_4 = 4 - 1x_2 .$$

Since x_3 and x_4 must be nonnegative it is clear that x_2 can be at most $\frac{7}{3}$. This is the smallest of the ratios $\theta_1 = \frac{7}{3}$, $\theta_2 = \frac{4}{1}$. The θ ratios are computed by dividing the entries in the right-hand column by the corresponding entries in the pivotal column. A pivotal row is then determined by choosing the smallest of the nonnegative ratios. In the interesting case of a tie for the smallest, either may be chosen. If we do not pick the smallest then some variable will become negative and the resulting solution will be infeasible. For example picking $\theta_2 = 4$ and increasing x_2 to 4 gives $x_3 = 7 - 12 = -5$. If an entry in the pivotal column is negative, the θ ratio will be negative, and the variable corresponding to that column may be increased without bound in that equation. For example suppose θ_1 were negative, then $x_3 = 7 + 3x_2$ and x_3 remains feasible for any increase in x_2. If an entry in the pivotal column is zero the θ ratio will be undefined and again no bound is placed on the corresponding x. Thus we consider θ only for the positive entries above some negative entry in the last row.

It is important to note that if all of the numbers above some negative number in the last row of a tableau are zero or negative, then the objective maximum is unbounded. Merely take the x of that column arbitrarily large. When this situation occurs we will say that the problem has an **unbounded optimal solution.**

We have now chosen a pivotal column, the second, and a pivotal row, the first. Our pivot is entry 3 circled in the initial tableau. Pivoting on 3 performs the Gauss-Jordan elimination to give the following tableau.

	x_1	x_2	x_3	x_4	
	$\frac{1}{3}$	1	$\frac{1}{3}$	0	$\frac{7}{3}$
	$\textcircled{\frac{5}{3}}$	0	$-\frac{1}{3}$	1	$\frac{5}{3}$
M	$-\frac{1}{3}$	0	$\frac{2}{3}$	0	$\frac{14}{3}$

This tableau shows that x_2 has come into the basis while x_3 went out. The new basic solution is $x_2 = \frac{7}{3}$, $x_4 = \frac{5}{3}$. We have increased x_2 from 0 to $\frac{7}{3}$ as predicted by the θ ratio, while M has increased from 0 to $\frac{14}{3}$. The presence of a negative in the last row indicates that we may pivot again, this time in the first column. Compute the θ ratios.

$$\theta_1 = \frac{7}{3} \cdot \frac{3}{1} = 7$$
$$\theta_2 = \frac{5}{3} \cdot \frac{3}{5} = 1$$

θ_2 is smaller and forces the pivotal row to be the second row. Carry out the pivoting on the entry $\frac{5}{3}$.

	x_1	x_2	x_3	x_4	
	0	1	$\frac{6}{15}$	$-\frac{1}{5}$	2
	1	0	$-\frac{1}{5}$	$\frac{3}{5}$	1
M	0	0	$\frac{9}{15}$	$\frac{1}{5}$	5

No negative entries in the last row indicate that M may not be increased further. We have the final tableau and read off the basic optimal feasible solution.

$$x_1 = 1$$
$$x_2 = 2$$
$$M = x_1 + 2x_2 = 1 + 4 = 5.$$

Theorem 3–1 assures us that no other feasible solution will have a larger value of M since no nonbasic feasible solution can have an objective value greater than that of all basic feasible solutions. ●

Definition 2. An **extended tableau** is **feasible** if it contains an identity matrix, and the numbers above the objective row in the right-hand column are nonnegative.

A feasible tableau contains a basic feasible solution. Likewise a tableau is called **infeasible**, or **unbounded**, or **optimal** if it contains respectively a basic infeasible, or unbounded, or optimal solution.

The method of choosing a pivot may be summarized in the following algorithm known as the Simplex Algorithm.

The Simplex Algorithm

1. Start with a tableau that is feasible. If there are no negative numbers in the bottom row then the tableau is already optimal. Otherwise proceed to Step 2.
2. Choose a pivotal column by picking that column among the first $n - 1$ columns whose entry in the bottom row is the most negative number.
3. For the first $m - 1$ rows compute the θ ratios for all positive entries in the pivotal column. (These are the ratios of the entry in the right-hand column to its corresponding entry in the pivotal column.) The θ ratios are all nonnegative.
4. Choose the pivotal row as one with the smallest θ ratio. In case of ties for the smallest, any one of the tying rows may be chosen.
5. Carry out a pivoting iteration using the pivot chosen by steps 2 and 4.
6. Repeat steps 2 through 5 until no new pivot can be found.

The algorithm will terminate when either no pivotal column or no pivotal row can be found. That is, either there are no negative coefficients in the objective row, meaning a basic optimal feasible solution has been found, or above some negative coefficient there are no positive entries in the pivotal column, meaning an unbounded optimal solution exists.

It is worthwhile to compare our basic feasible solutions obtained by the simplex method with the graphical solution. The feasible region of the original problem is the intersection of four half planes as shown in Figure 3–1. The vertices of the convex polygon are $(0, 0)$, $(2, 0)$, $(1, 2)$ and $(0, \frac{7}{3})$.

After slack variables are introduced the solution space is four-dimensional. The equations in our tableaux are three-dimensional hyperplanes in four-space. The solution points are in the common two-dimensional space of these hyperplanes. If we project the solution set (x_1, x_2, x_3, x_4) onto the linear subspace $(x_1, x_2, 0, 0)$ and associate this subspace with the two-space (x_1, x_2), then the set of feasible solutions can be considered to be the same as the convex set in Figure 3–1. The fact that the slack variables have zero coefficients in the objective function assures us that a solution will be projected onto $(x_1, x_2, 0, 0)$

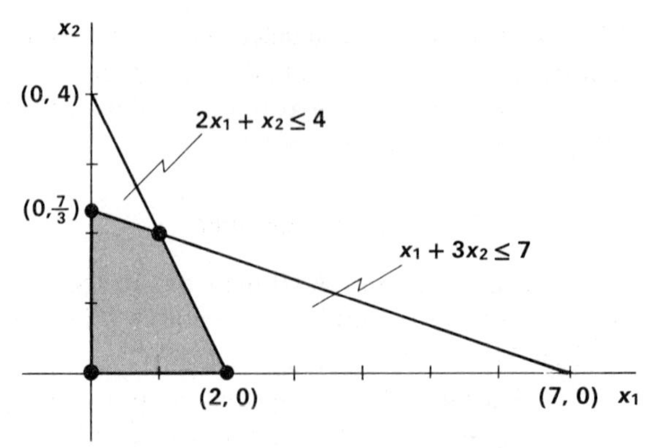

FIGURE 3–1

and is thus associated with a point (x_1, x_2). Making this association we have the following solution points in our tableaux:

	$(x_1, x_2, x_3, x_4) \to (x_1, x_2)$
first B.F.S.	$(0, 0, 7, 4) \to (0, 0)$
second B.F.S.	$(0, \frac{7}{3}, 0, \frac{5}{3}) \to (0, \frac{7}{3})$
third B.F.S.	$(1, 2, 0, 0) \to (1, 2)$.

Note in our figure that the process began at a vertex, the origin, and proceeded to adjacent vertices until the optimal vertex was reached. The simplex method is merely a way to move from one B.F.S. (basic feasible solution) to another with a value of M at least as great. Each move is equivalent to going from one vertex to an adjacent vertex of the feasible region. Since there are only a finite number of vertices, the optimal vertex must be reached in a finite number of steps provided we improve the objective value at each step.

Problems Section 3.2

For the first three problems solve the given set of linear equations by pivoting down the main diagonal.

3–1.
$$\begin{aligned} 3x - y + 6z &= 1 \\ x + 2y - 3z &= 0 \\ 2x - 3y - z &= -9 \end{aligned}$$

3–2.
$$\begin{aligned} 5x - y + 3z &= 1 \\ 2x + 3y - z &= -1 \\ 10x + 4y + 3z &= -4 \end{aligned}$$

3–3. $$\begin{aligned} 2x - 3y - z &= 11 \\ x + 4y - 2z &= 3 \\ 3x + 4y + 5z &= -5 \end{aligned}$$

Solve the following linear programming problems by pivoting the extended tableau.

3–4. Find nonnegative numbers x_1 and x_2 satisfying the constraints

$$\begin{aligned} x_1 + 2x_2 &\leq 10 \\ 3x_1 + 5x_2 &\leq 27, \end{aligned}$$

such that $11x_1 + 20x_2$ is a maximum. Check your answer graphically.

3–5. Find nonnegative numbers x_1 and x_2 satisfying the constraints

$$\begin{aligned} x_1 + 2x_2 &\leq 8 \\ 2x_1 + 3x_2 &\leq 13 \\ x_1 + x_2 &\leq 6, \end{aligned}$$

such that $8x_1 + 9x_2$ is a maximum. Check your answer graphically.

3–6. Solve Problem 1–14 finding the maximum profit of Seeall's Manufacturing Company.

3–7. Find nonnegative numbers x_1, x_2, and x_3 satisfying the constraints

$$\begin{aligned} 3x_1 - 4x_2 + 8x_3 &\leq 10 \\ 4x_1 - 2x_2 &\leq 12 \\ -x_1 + 3x_2 + 2x_3 &\leq 7, \end{aligned}$$

such that $3x_1 - x_2 - 2x_3$ is a maximum.

3–8. Caroline's Quality Candy Confectionery is famous for fudge, chocolate cremes, and pralines. Their candy making equipment is set up to make 100 lb. batches at a time. Currently there is a chocolate shortage and they can get only 120 lbs. of chocolate in the next shipment. On a week's run the confectionery's cooking and processing equipment is available for a total of 42 machine hours. During the same period the employees have a total of 56 man-hours available for packaging. A batch of fudge requires 20 lbs. of chocolate while a batch of cremes uses 25 lbs. of chocolate. The cooking and processing takes 120 minutes for fudge, 150 minutes for chocolate cremes, and 200 minutes for pralines. The packaging times measured in minutes per one pound box are 1, 2, and 3 respectively for fudge, cremes, and pralines. Determine how many batches of each type of candy the confectionary should make assuming that the profit per pound box is 50 cents on fudge, 40 cents on chocolate cremes, and 45 cents on pralines. Also find the maximum profit for the week.

3.3 The Condensed Tableau

Example 3. Consider again Example 2 on page 29. Find $x_1 \geq 0$, $x_2 \geq 0$ such that

$$x_1 + 3x_2 \leq 7$$
$$2x_1 + x_2 \leq 4$$

and $x_1 + 2x_2 = M$ is a maximum. You may note that the columns of unit vectors in the extended tableau offered no significant information other than to designate the basic variables. It will be convenient to omit these columns and place the subscripts of the basic variables in a column to the left of our tableau. These subscripts will indicate the row of that basic variable and its corresponding basic value will be found in the right-hand column. Finally the columns will be labeled at the top of our tableau with the subscripts of the corresponding non-basic variables. The initial **condensed tableau** that appears below contains all of the same information as the corresponding extended tableau while omitting its unit column vectors.

	1	2	
3	1	(3)	7
4	2	1	4
M	−1	−2	0

Notice how easily this tableau may be set up from the original constraints and the basic solution read off, $x_3 = 7$, $x_4 = 4$, and $M = 0$. As before, the pivot is determined by the θ ratios of the second column, $\theta_1 = \frac{7}{3}$, $\theta_2 = \frac{4}{1}$, to be the number 3 circled above. Now look at the second tableau of the previous section and condense it into the following.

	1	3	
2	$\frac{1}{3}$	$\frac{1}{3}$	$\frac{7}{3}$
4	$\frac{5}{3}$	$-\frac{1}{3}$	$\frac{5}{3}$
M	$-\frac{1}{3}$	$\frac{2}{3}$	$\frac{14}{3}$

This tableau contains all of the necessary information, and the second B.F.S. is read off immediately as $x_2 = \frac{7}{3}$, $x_4 = \frac{5}{3}$, $M = \frac{14}{3}$.

The secret is to determine this second tableau directly from the first condensed tableau without resorting to the extended tableau. A close comparison of the two condensed tableaux reveals the following five step algorithm.

The Condensed Tableau Pivoting Algorithm

1. Replace the pivot with its reciprocal.
2. Divide the remaining entries of the pivotal row by the pivot.
3. Divide the remaining entries of the pivotal column by the negative of the pivot.
4. For the entries in the remaining positions subtract from the corresponding position in the old tableau the product of the old value both in that row and in the pivotal column times the new entry both in that column and in the pivotal row.
5. Interchange the integer above the pivotal column with the integer to the left of the pivotal row.

Try these five steps out on the second tableau, pivoting on the number $\frac{5}{3}$, to check the final tableau.

	1	3	
2	$\frac{1}{3}$	$\frac{1}{3}$	$\frac{7}{3}$
4	$\left(\frac{5}{3}\right)$	$-\frac{1}{3}$	$\frac{5}{3}$
M	$-\frac{1}{3}$	$\frac{2}{3}$	$\frac{14}{3}$

$\rightarrow$

	4	3	
2	$-\frac{1}{5}$	$\frac{6}{15}$	2
1	$\frac{3}{5}$	$-\frac{1}{5}$	1
M	$\frac{1}{5}$	$\frac{9}{15}$	5

To illustrate step 4 suppose the matrix entries are labeled Y_{ij}, i running from 1 to $m = 3$, j running from 1 to $n = 3$. Let the pivotal row be p and the pivotal column be q so that the pivot is Y_{pq}. Then any new entry $\overline{Y}_{ij}$ that is neither in the pivotal row nor in the pivotal column is found according to step 4 by the formula

$$\overline{Y}_{ij} = Y_{ij} - Y_{iq}\frac{Y_{pj}}{Y_{pq}}$$

To compute the final value of M we have

$$\overline{Y}_{33} = Y_{33} - Y_{31}\frac{Y_{23}}{Y_{21}} = \frac{14}{3} - \left(-\frac{1}{3}\right)\frac{\frac{5}{3}}{\frac{5}{3}} = 5. \quad \bullet$$

The whole process is readily programmed for a digital computer and such a program is given in the next chapter.

A proof that the steps of our algorithm hold is accomplished by condensing the tableaux of a pivot of the general extended tableau and then comparing these condensed tableaux. The general extended tableau is indicated schematically below where variable s is chosen to come into the basis replacing basic variable u. The pivot is Y_{pq}.

		pivotal column s		some other column		u		
		$\vdots$		$\vdots$		$\vdots$		$\vdots$
pivotal row	$\cdots$	Y_{pq}	$\cdots$	Y_{pj}	$\cdots$	1	$\cdots$	Y_{pn}
		$\vdots$		$\vdots$		$\vdots$		$\vdots$
some other row	$\cdots$	Y_{iq}	$\cdots$	Y_{ij}	$\cdots$	0	$\cdots$	Y_{in}
		$\vdots$		$\vdots$		$\vdots$		$\vdots$

The extended tableau pivoting process produces the following tableau where s is now basic.

	s				u		
	$\vdots$		$\vdots$		$\vdots$		$\vdots$
$\cdots$	1	$\cdots$	Y_{pj}/Y_{pq}	$\cdots$	$1/Y_{pq}$	$\cdots$	Y_{pn}/Y_{pq}
	$\vdots$		$\vdots$		$\vdots$		$\vdots$
$\cdots$	0	$\cdots$	$Y_{ij} - Y_{iq}(Y_{pj}/Y_{pq})$	$\cdots$	$-Y_{iq}/Y_{pq}$	$\cdots$	$Y_{in} - Y_{iq}(Y_{pn}/Y_{pq})$
	$\vdots$		$\vdots$		$\vdots$		$\vdots$

Condense the first tableau by dropping all columns of unit vectors. Note u appears to the left opposite its value Y_{pn}.

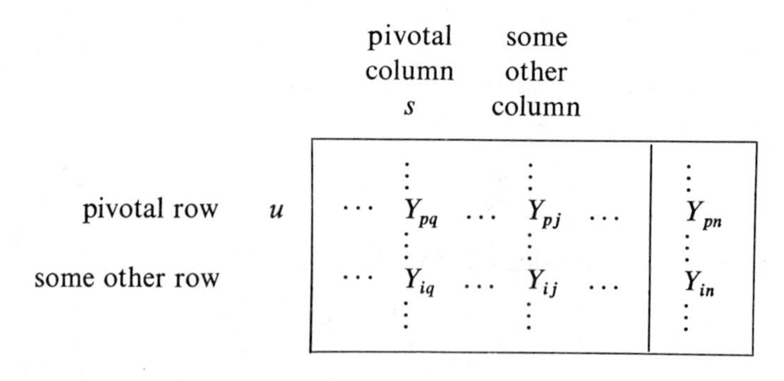

			pivotal column s		some other column		
			$\vdots$		$\vdots$		$\vdots$
pivotal row	u	$\cdots$	Y_{pq}	$\cdots$	Y_{pj}	$\cdots$	Y_{pn}
			$\vdots$		$\vdots$		$\vdots$
some other row		$\cdots$	Y_{iq}	$\cdots$	Y_{ij}	$\cdots$	Y_{in}
			$\vdots$		$\vdots$		$\vdots$

In condensing the second extended tableau replace the s column with the u column and drop all other columns of unit vectors.

		u				
		⋮		⋮		⋮
s	⋯	$1/Y_{pq}$	⋯	Y_{pj}/Y_{pq}	⋯	Y_{pn}/Y_{pq}
		⋮		⋮		⋮
	⋯	$-Y_{iq}/Y_{pq}$	⋯	$Y_{ij} - Y_{iq}(Y_{pj}/Y_{pq})$	⋯	$Y_{in} - Y_{iq}(Y_{pn}/Y_{pq})$
		⋮		⋮		⋮

The five step algorithm is clearly evident by comparing these condensed tableaux.

The Simplex Algorithm for choosing a pivot applies as well to the condensed tableau as it did to the extended tableau. The only difference is that the pivoting is now carried out by the Condensed Tableau Pivoting Algorithm.

Problems Section 3.3

Solve the following linear programs by pivoting the condensed tableau.

3–9. Find $x_1 \geq 0$ and $x_2 \geq 0$ such that

$$\begin{aligned} 2x_1 + 3x_2 &\leq 30 \\ x_1 + x_2 &\leq 11 \\ 4x_1 + 3x_2 &\leq 40, \end{aligned}$$

and $5x_1 + 6x_2$ is a maximum.

3–10. Find $x_1 \geq 0$ and $x_2 \geq 0$ such that

$$\begin{aligned} 2x_1 + 3x_2 &\leq 30 \\ x_1 + x_2 &\leq 11 \\ 4x_1 + 3x_2 &\leq 40, \end{aligned}$$

and $7x_1 + 6x_2$ is a maximum. Check your answer graphically.

3–11. Find $x_1 \geq 0$, $x_2 \geq 0$, and $x_3 \geq 0$ such that

$$\begin{aligned} x_1 + x_2 + 2x_3 &\leq 200 \\ 2x_1 + x_2 &\leq 100 \\ 10x_1 + 8x_2 + 5x_3 &\leq 2000, \end{aligned}$$

and $2x_1 + 4x_2 + x_3$ is a maximum.

3–12. Find $x_1 \geq 0$ and $x_2 \geq 0$ such that

$$\begin{aligned} 5x_1 + 2x_2 &\leq 45 \\ -4x_1 + 5x_2 &\leq 30, \end{aligned}$$

and $-2x_1 + 5x_2$ is a maximum. Find the minimum of the same objective function. Hint: Let the minimum be the negative of a maximum, $m = -M$, and then maximize the expression M. Check your answers graphically.

3–13. Find $x_1 \geq 0$ and $x_2 \geq 0$ such that

$$\begin{aligned} x_1 - x_2 &\geq -3 \\ 3x_1 + 4x_2 &\leq 40 \\ x_1 &\leq 9 \\ 7x_1 + 2x_2 &\leq 64, \end{aligned}$$

and $4x_1 + x_2$ is a maximum. Hint: How can you reverse the sense of the first inequality?

3–14. With the same constraints of Problem 3–13 maximize $3x_1 + x_2$.

3–15. With the same constraints of Problem 3–13 maximize $x_1 - 2x_2$.

3–16. With the same constraints of Problem 3–13 minimize $x_1 - 2x_2$. Note the hint in Problem 3–12.

3–17. A small boat manufacturer builds three types of fiberglass fishing boats, a pram, a run-a-bout, and a trihull whaler. He sells the pram at a profit of \$75, the run-a-bout at a profit of \$90, and the whaler at a profit of \$100. The factory is divided into two sections. Section A does the molding and construction work while section B does the painting, finishing, and equipping. The pram takes 1 hour in section A and 2 hours in B. The run-a-bout takes 2 hours in section A and 5 hours in B. The whaler takes 3 hours in section A and 4 hours in B. Shop A has a total of 6240 hours available and shop B has a total of 10,800 hours available for the year. The manufacturer has ordered a year's supply of fiberglass that is sufficient to build at most 3000 boats figuring the average amount of fiberglass used per boat. What is his year's production schedule and profit in order to earn the maximum profit?

3–18. Suppose in Problem 3–17 that the run-a-bout sold for a profit of \$100 and the whaler sold for a profit of \$90 while the profit on the pram remained at \$75. Then what should the production schedule and maximum profit be?

3.4 Artificial Variables

So far in Chapter 3 we have considered only type I inequalities ($\leq$). A linear program may have constraints utilizing type II inequalities ($\geq$) or even equalities. In the latter two cases artificial variables are introduced in order to find an initial basic feasible solution. Let us consider the following problem which illustrates all three types of constraints.

Example 4. The Navy is experimenting with three types of bombs, A, B, and C, in which there will be used three kinds of explosives, D, E, and F. The Captain wishes to use exactly 2000 pounds of explosive D, at least 1000 pounds of explosive E, and at most 3000 pounds of explosive F. Bomb A requires 3, 2, 1 pounds of D, E, and F respectively. Bomb B requires 1, 5, 2 pounds of D, E, and F respectively. Bomb C requires 6, 1, 4 pounds of D, E, and F respectively. Now bomb A will give the equivalent of a 1 ton explosion, bomb B will give a 4 ton explosion, and bomb C will give a 3 ton explosion. Under what production schedule can the Navy make the biggest bang?

Solution. First, choose the unknowns x_1, x_2, and x_3 for the number of bombs to be produced of types A, B, and C respectively. The constraints on the amount of each kind of explosive available give the following system

$$\langle \mathbf{1} \rangle \qquad \begin{aligned} 3x_1 + 1x_2 + 6x_3 &= 2000 \\ 2x_1 + 5x_2 + 1x_3 &\geq 1000 \\ 1x_1 + 2x_2 + 4x_3 &\leq 3000. \end{aligned}$$

The objective is to create the largest explosive force, so

$$1x_1 + 4x_2 + 3x_3 = M$$

should be maximized subject to constraints $\langle \mathbf{1} \rangle$ and the nonnegativity requirement $x_1 \geq 0$, $x_2 \geq 0$, $x_3 \geq 0$.

Let us set up the extended tableau and then condense it. We introduce slack variables x_4 and x_7 to convert the inequalities into equalities. System $\langle \mathbf{1} \rangle$ now appears as

$$\langle \mathbf{2} \rangle \qquad \begin{aligned} 3x_1 + 1x_2 + 6x_3 &= 2000 \\ 2x_1 + 5x_2 + 1x_3 - 1x_4 &= 1000 \\ 1x_1 + 2x_2 + 4x_3 + 1x_7 &= 3000. \end{aligned}$$

The requirement that the slack variables also be nonnegative forces the coefficient of x_4 to be negative. In this case the slack variable represents an excess that must be subtracted from the left side of the inequality to reduce it to equality. Unfortunately we have only one column unit vector and no basic feasible solution. In order to gain two more column unit vectors we introduce two *artificial variables* $x_{-5} \geq 0$ and $x_{-6} \geq 0$ into the first two equations of $\langle \mathbf{2} \rangle$. The subscripts are made negative so that a computer may easily recognize artificial variables. To keep from losing track of the original problem while solving the *artificial problem*, we charge an arbitrarily high cost for the artificial variables in the objective function. Let the objective function of the artificial problem be

$$\overline{M} = M - N(x_{-5} + x_{-6})$$

where N is a large positive number. Since an arbitrarily large amount is subtracted from M, $\overline{M}$ cannot possibly reach a maximum unless $x_{-5} = x_{-6} = 0$. Due to this high cost N, pivoting will drive the artificial variables out of the basis forcing them to zero and leaving $\overline{M} = M$. The artificial problem can be stated as follows: Find x_1, x_2, x_3, x_4, x_{-5}, x_{-6}, x_7 all ≥ 0 such that

$$\begin{aligned} 3x_1 + 1x_2 + 6x_3 + 1x_{-5} &= 2000 \\ 2x_1 + 5x_2 + 1x_3 - 1x_4 \quad + 1x_{-6} &= 1000 \\ 1x_1 + 2x_2 + 4x_3 + 1x_7 \quad &= 3000, \end{aligned}$$

and $1x_1 + 4x_2 + 3x_3 - N(x_{-5} + x_{-6}) = \overline{M}$ is a maximum.

In order to separate the arbitrarily large coefficients from the relatively small coefficients, create a double row for the objective function where the bottom row of the tableau represents the multiples of N. Here is the initial extended tableau.

	x_1	x_2	x_3	x_4	x_{-5}	x_{-6}	x_7	
	3	1	6	0	1	0	0	2000
	2	5	1	−1	0	1	0	1000
	1	2	4	0	0	0	1	3000
$\overline{M}$	−1	−4	−3	0	0	0	0	0
	0	0	0	0	1	1	0	0

The ones must now be removed from the bottom row in order that x_{-5}, x_{-6}, x_7 and $\overline{M}$ will form a basic solution. This is easily done by multiplying each of the first two equations by N and subtracting from the objective equation.

	x_1	x_2	x_3	x_4	x_{-5}	x_{-6}	x_7	
	3	1	6	0	1	0	0	2000
	2	5	1	−1	0	1	0	1000
	1	2	4	0	0	0	1	3000
$\overline{M}$	−1	−4	−3	0	0	0	0	0
	−5	−6	−7	1	0	0	0	−3000

The first basic feasible solution is read off as $x_{-5} = 2000$, $x_{-6} = 1000$, and $x_7 = 3000$ while the remaining variables are zero. The starting value of $\overline{M}$ is a large negative number, $-3000N$. Dropping the column unit vectors we have the following initial condensed tableau.

	1	2	3	4	
−5	3	1	(6)	0	2000
−6	2	5	1	−1	1000
7	1	2	4	0	3000
$\overline{M}$	−1	−4	−3	0	0
	−5	−6	−7	1	−3000

The pivoting is carried out by choosing a pivotal column according to negatives in the last row as long as artificial variables remain in the basis. As soon as an artificial variable is removed from the basis, that column may be dropped from the tableau. Since a maximum for $\overline{M}$, if it exists, requires that all artificial variables go out of the basis, we will never pivot again on a column headed by such a variable. Checking the θ ratios for the third column, above entry -7, we find entry 6 for the first pivot. The resulting tableaux listed below were found on a computer.

	1	2	−5	4	
3	.5	.167	.167	0	333.33
−6	1.5	(4.83)	−.167	−1	666.67
7	−1	1.33	−.667	0	1666.67
$\overline{M}$	.5	−3.5	.5	0	1000.
	−1.5	−4.83	1.167	1	−666.67

The third column is dropped, and the second pivot is found to be entry 4.83 to obtain the next tableau. The artificial variables are now out of the basis and may be neglected. Thus, the second column may be dropped. Notice that the remaining entries in the last row are all zeros and in particular $\overline{M} = M$ is no longer affected by the last row. Dropping the last row completely we next enter phase two of a two phase

	1	−6	4	
3	.448	−.034	.034	310.35
2	.310	.207	−.207	137.93
7	−1.414	−.276	(.276)	1482.76
$\overline{M}$	1.586	.724	−.724	1482.76
	0	1	0	0

problem. Eliminating the artificial variables is often called phase one and then the second phase continues the pivoting according to negative entries in the first of the two objective rows. Our third pivot is thus found above −.724 to be .276.

	1	7	
3	(.625)	−.125	125.
2	−.750	.750	1250.
4	−5.125	3.625	5375.
M	−2.125	2.625	5375.

The fourth pivot must be .625 because that is the only positive entry above the negative entry −2.125. This final pivot gives the optimal tableau listed below.

	3	7	
1	1.6	−.2	200
2	1.2	.6	1400
4	8.2	2.6	6400
M	3.4	2.2	5800

No further pivoting is possible because the last row is all positive. The final solution is read off to be $x_1 = 200$, $x_2 = 1400$, $x_3 = 0$, $x_4 = 6400$,

$x_7 = 0$, and $M = 5800$. The Navy should produce 200 bombs of type A, 1400 bombs of type B, and none of the type C bombs for a total explosive power of 5800 tons. Since slack variable $x_7 = 0$, all 3000 pounds of explosive F will be used. Slack variable $x_4 = 6400$ means that an additional 6400 pounds of explosive E is utilized above the required 1000 pounds. These results may all be checked in the original constraints. ●

3.5 Minimization

The previous examples in this chapter have called for maximizing the objective function. It is convenient to convert every linear programming problem into the case of finding a maximum so that one computer routine will handle both cases. This may easily be done. If the problem calls for minimizing the objective function then maximize the negative of the objective function. This essentially reflects the problem with respect to the origin and the desired minimum is the negative of the maximum found in the final tableau.

Example 5. Consider Example 4 on page 41 with a new objective. Under the same constraints on the amounts of explosive to be used, let us find the production schedule that will give the minimum cost of explosive ingredients. Suppose explosive D costs \$100 per lb., explosive E costs \$500 per lb., and explosive F costs \$300 per lb. Find the minimum cost.

Solution. Select two slack variables, x_4, x_7, and two artificial variables, x_{-5}, x_{-6}. The slack variables have zero cost and the artificial variables have a very large cost, N. The three types of bombs, A, B, and C, will cost \$1600, \$3200, and \$2300 respectively. The objective cost equation is thus

$$\langle 1 \rangle \qquad m = 1600x_1 + 3200x_2 + 2300x_3 .$$

First set $m = -M$ where M is to be maximized. Then set up the artificial problem to maximize $\overline{M}$ where

$$\langle 2 \rangle \qquad \overline{M} = M - N(x_{-5} + x_{-6}).$$

By combining $\langle 1 \rangle$ and $\langle 2 \rangle$ the objective is

$$\langle 3 \rangle \qquad \overline{M} + 1600x_1 + 3200x_2 + 2300x_3 + N(x_{-5} + x_{-6}) = 0.$$

The initial extended tableau is given below.

	x_1	x_2	x_3	x_4	x_{-5}	x_{-6}	x_7	
	3	1	6	0	1	0	0	2000
	2	5	1	−1	0	1	0	1000
	1	2	4	0	0	0	1	3000
$\overline{M}$	1600	3200	2300	0	0	0	0	0
	0	0	0	0	1	1	0	0

Once again we must eliminate the 1's in the last row by subtracting N times the first two rows, giving

	x_1	x_2	x_3	x_4	x_{-5}	x_{-6}	x_7	
	3	1	6	0	1	0	0	2000
	2	5	1	−1	0	1	0	1000
	1	2	4	0	0	0	1	3000
$\overline{M}$	1600	3200	2300	0	0	0	0	0
	−5	−6	−7	1	0	0	0	−3000

Eliminating unit vectors, the initial condensed tableau is

	1	2	3	4	
−5	3	1	⑥	0	2000
−6	2	5	1	−1	1000
7	1	2	4	0	3000
$\overline{M}$	1600	3200	2300	0	0
	−5	−6	−7	1	−3000

With a little skill and practice one may set up the condensed tableau directly without resorting to the extended tableau.

The first two pivots occur at the same entries as in the previous problem. The pivots are circled in each tableau. Every time an artificial variable is removed from the basis that column is dropped. The problem is now solved on a computer that lists the following sequence of three tableaux.

	1	2	−5	4	
3	.5	.167	.167	0	333.33
−6	1.5	(4.833)	−.167	−1	666.67
7	−1.	1.333	−.667	0	1666.67
$\overline{M}$	450.	2817.	−383.	0	−766667.
	−1.5	−4.833	1.167	1	−666.67

	1	−6	4	
3	.448	−.0345	.0345	310.35
2	(.310)	.207	−.207	137.93
7	−1.414	−.276	.276	1482.76
$\overline{M}$	−424.1	−582.8	582.8	−1155170.
	0	1	0	0

	2	4	
3	−1.444	.333	111.11
1	3.222	−.667	444.44
7	4.556	−.667	2111.11
M	1366.7	300.	−966667.

Our problem is completed on three pivots and the answer read off from the final tableau. As frequently happens the answers are not integral. In general, the difficulty may be resolved by Gomory's Algorithm that is discussed in Chapter 8. At this time we accept the closest integer as an approximate answer. The Navy should make 444 of the type A bombs, none of the type B bombs, and 111 of the type C bombs at a minimum cost of \$966,667. Since slack variable $x_4 = 0$, only 1000 pounds of explosive E are used. The slack $x_7 = 2111$ indicates that 2111 pounds of the 3000 pounds of explosive F remains unused. If these results are checked in the original constraints the discrepancies due to our rounding off will show up. ●

3.6 Unboundedness and Inconsistency

It has been mentioned that a linear programming problem may not have a solution. This situation should be detectable from the final tableau. The pivoting procedure continues to produce basic feasible solutions until no new pivot can be found. The process might stop because no pivotal row can be found to go with a possible pivotal column. This case occurs when there are no positive entries above some negative entry in the last row. The variable corresponding to this column could then be brought into the basis at an arbitrarily large value causing the objective function to be unbounded. The situation is easily noted in the final tableau as illustrated in the next example.

Example 6. Find nonnegative numbers x_1 and x_2 such that

$$x_1 - 2x_2 \leq 2$$
$$3x_1 - 2x_2 \leq 18,$$

and $2x_1 - x_2$ is a maximum. The initial tableau and two pivots are given below,

	1	2	
3	①	-2	2
4	3	-2	18
M	-2	1	0

Pivoting at $p, q = 1, 1$:

	3	2	
1	1	-2	2
4	-3	④	12
M	2	-3	4

Pivoting at $p, q = 2, 2$:

	3	4	
1	$-\frac{1}{2}$	$\frac{1}{2}$	8
2	$-\frac{3}{4}$	$\frac{1}{4}$	3
M	$-\frac{1}{4}$	$\frac{3}{4}$	13

We have arrived at the final tableau since no pivotal row can be found. However a negative in the objective row indicates that a maximum has not been reached. If we were to bring x_3 back into the basis the equations which should limit the value of x_3 are

$$\langle \mathbf{1} \rangle \qquad \begin{aligned} x_1 &= 8 + \tfrac{1}{2}x_3 \\ x_2 &= 3 + \tfrac{3}{4}x_3 . \end{aligned}$$

The value of x_3 may be arbitrarily large in equations $\langle \mathbf{1} \rangle$ without destroying the feasibility of x_1 and x_2. Consequently x_1 and x_2 may be arbitrarily large subject only to

$$x_1 = \tfrac{2}{3}x_2 + 6.$$

The result is clearly illustrated in Figure 3–2.

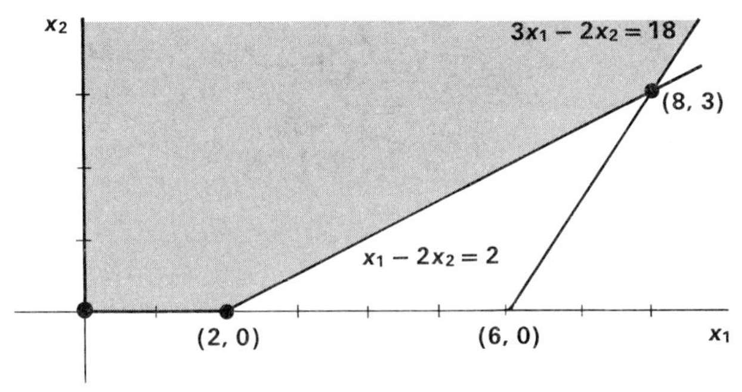

FIGURE 3–2

The feasible region is the unbounded shaded area in Figure 3–2. The three basic feasible solutions found in the above tableaux are respectively the vertices (0, 0) (2, 0), and (8, 3). By continuing from vertex (8, 3) upward along the boundary $3x_1 - 2x_2 = 18$, the objective function may be increased without bound. If $x_1 = 100$ then $x_2 = 141$, and the objective is $M = 2(100) - 141 = 59$. The final tableau must be viewed carefully to see if a maximum has truly been reached. ●

Another way in which a solution may not exist is in the possibility that the original constraints are inconsistent. This happens when an artificial variable cannot be removed from the basis. A basic artificial variable in the final tableau signifies that we cannot find a basic feasible solution to the original problem. The artificial variables were introduced for the sole purpose of finding an initial basic feasible solution to the original problem and they must go out to get such a solution. Thus, if an artificial variable remains in the final tableau, there are no feasible points and the original constraints must be inconsistent. Consider the following example:

Example 7. Find $x_1 \geq 0$ and $x_2 \geq 0$ such that

$$2x_1 + 5x_2 \leq 10$$
$$-4x_1 + 7x_2 \geq 21,$$

and $x_1 + 2x_2$ is a maximum. In order to set up the initial tableau let x_3 and x_4 be the slack variables, and let x_{-5} be the artificial variable for the type II constraint. Let $M = x_1 + 2x_2$ and let the maximum of the artificial problem be $\overline{M} = M - Nx_{-5}$. Then the new problem appears as follows. Find $x_1, x_2, x_3, x_4, x_{-5}$ all ≥ 0 such that

$$2x_1 + 5x_2 + x_3 = 10$$
$$-4x_1 + 7x_2 - x_4 + x_{-5} = 21,$$

and $\overline{M} = x_1 + 2x_2 - Nx_{-5}$ is a maximum.

The initial condensed tableau with a double row for the objective function is given below. Notice that in order to get x_{-5} into the basis the second row has been subtracted from the last row.

	1	2	4	
3	2	⑤	0	10
−5	−4	7	−1	21
$\overline{M}$	−1	−2	0	0
	4	−7	1	−21

Pivoting on 5 at matrix position (1, 2) produces the next and final tableau.

	1	3	4	
2	.4	.2	0	2
−5	−6.8	−1.4	−1	7
$\overline{M}$	−.2	.4	0	4
	6.8	1.4	1	−7

Since there are no negative numbers in the last row, there is no possible pivotal column. The final basic solution is $x_2 = 2$, $x_{-5} = 7$. $\overline{M} = 4 - 7N$ where N is arbitrarily large. Thus $\overline{M}$ is an arbitrary negative number and never reaches a maximum. The artificial problem never reduces to the original problem. Therefore, the original problem has no solution and its constraints must be inconsistent. The inconsistency shows up on the following graph, Figure 3–3. It shows that there are no feasible points. ●

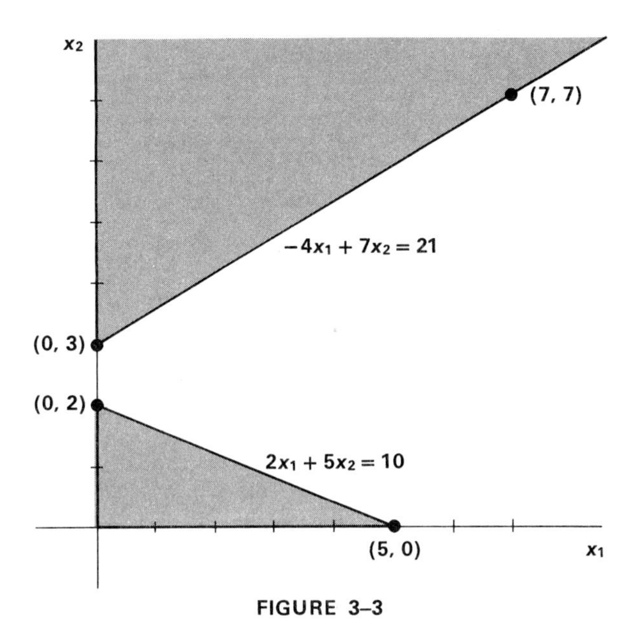

FIGURE 3–3

3.7 Degeneracy

Another difficulty in the simplex method arises when one of the basic variables is zero in addition to the nonbasic variables that are zero by definition. If a basic variable becomes zero in a given tableau, then one of the θ ratios used for finding the next pivotal row may be zero. Since zero is the minimum possible θ ratio, the next pivotal row would be that one with the zero basic variable. The value of the minimum θ ratio is also the value of the new variable introduced into the basis. By step 2 of our algorithm for pivoting the condensed tableau, this new variable comes into the basis at the old value divided by the pivot. Thus the new variable introduced will remain at value zero. By checking step 4 of the algorithm

$$\overline{Y}_{in} = Y_{in} - Y_{iq}\, Y_{pn}/Y_{pq},$$

we see that when $Y_{pn} = 0$, Y_{in} remains unchanged for all i. This means that all basic variables in the new tableau have the same value that they held in the old tableau. The only difference in the final column of the two tableaux is that the basic variable with value zero has been renamed. In particular the objective value has not changed. We will call such a basic feasible solution degenerate.

Definition 3. A basic feasible solution that has one or more basic variables equal to zero is called **degenerate.**

Theorem 3–2. *Degeneracy will occur whenever there is a tie for least among the θ ratios used to determine a pivotal row.*

Proof. Consider the condensed tableau,

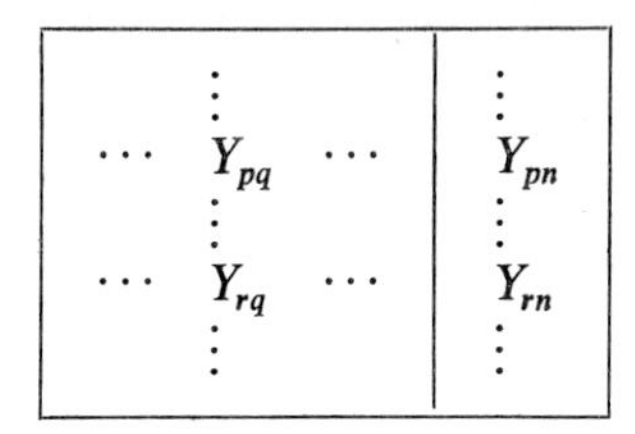

and suppose there is a tie between the pth and r^{th} rows for the least θ ratio.

Then,

$$Y_{pn}/Y_{pq} = Y_{rn}/Y_{rq}. \tag{1}$$

Choose Y_{pq} as pivot and carry out the pivoting algorithm.

	$\vdots$		$\vdots$
$\cdots$	$1/Y_{pq}$	$\cdots$	Y_{pn}/Y_{pq}
	$\vdots$		$\vdots$
$\cdots$	$-Y_{rq}/Y_{pq}$	$\cdots$	$Y_{rn} - Y_{rq}(Y_{pn}/Y_{pq})$
	$\vdots$		$\vdots$

From equation ⟨1⟩,

$$Y_{rn} - Y_{rq}\,Y_{pn}/Y_{pq} = 0.$$

Thus, a basic variable is zero and the corresponding basic feasible solution is degenerate. ■

In a practical case degeneracy does not cause any real trouble. Pivoting continues until an optimum tableau is reached. However, the optimum solution found may not be unique. The fact that the objective value may remain constant through several pivots means alternate optima are possible. There is a theoretical difficulty. If the objective value cannot be improved at each pivot then we cannot be sure that a final answer will ever be reached. On the other hand, as long as the objective is improved with each pivot we must reach the optimum in a finite number of steps since there are only a finite number of basic solutions. In the degenerate case it is possible to get into a cycle while the objective function is constant. This means that after a number of pivots you return to a previously encountered basic solution. Of course this would be disastrous on a computer. The machine would be in an infinite loop with no way to get out. While this does not occur in practice, examples have been constructed to illustrate the point.

The following example is due to E. M. L. Beale:[3]

Example 8.

	1	2	3	4	
5	$\frac{1}{4}$	-8	-1	9	0
6	$\frac{1}{2}$	-12	$-\frac{1}{2}$	3	0
7	0	0	1	0	1
	$-\frac{3}{4}$	20	$-\frac{1}{2}$	6	0

If all ties in this example are broken by choosing the basic variable of lowest subscript to go out of the basis, then there is a cycle of period six. The basic variables in the cycle are: (5, 6, 7), (1, 6, 7), (1, 2, 7), (3, 2, 7), (3, 4, 7), (5, 4, 7), (5, 6, 7). To get out of the cycle all that is necessary is to break the ties in some other order. It is possible to write a computer program that will break all ties in such a way to make cycling impossible. The technique is called perturbation. It consists of changing the last column slightly so that ties do not occur. ●

Perturbation is discussed in Vajda's *Mathematical Programming*, Section 5–3. The programs to be presented next in Chapter 4 do not bother with the possibility of cycling.

Problems Section 3.4–3.7

The following problems should be done by pivoting the condensed tableau on a computer. Programming aids are given in Chapter 4. At first the

[3] E. M. L. Beale, "Cycling in the Dual Simplex Algorithm," *Naval Research Logistics Quarterly*, 2, 1955: 269–275.

pivoting may be done one tableau at a time by telling the machine the pivot position with an input command. Eventually a completely automatic program should be compiled so that the machine will find the pivot position and continue to iterate until the final tableau is reached. Only the final tableau need be printed out. Add artificial variables where necessary and note the cases of unboundedness, inconsistancy or degeneracy.

3–19. Find nonnegative numbers x_1, x_2, x_3 and the maximum value of M for

$$\begin{aligned} x_1 + 2x_2 - x_3 &\leq 9 \\ 2x_1 - x_2 + 2x_3 &= 4 \\ -x_1 + 2x_2 + 2x_3 &\geq 5 \\ 2x_1 + 4x_2 + x_3 &= M. \end{aligned}$$

3–20. Find nonnegative numbers x_1, x_2, x_3 and the minimum value of m for

$$\begin{aligned} x_1 + 2x_2 - x_3 &\leq 9 \\ 2x_1 - x_2 + 2x_3 &= 4 \\ -x_1 + 2x_2 + 2x_3 &\geq 5 \\ 2x_1 + 4x_2 + x_3 &= m. \end{aligned}$$

3–21. Find nonnegative numbers x_1, x_2, x_3 and the minimum value of m for

$$\begin{aligned} x_1 + 2x_2 - x_3 &\leq 5 \\ 2x_1 - x_2 + 2x_3 &= 2 \\ -x_1 + 2x_2 + 2x_3 &\geq 1 \\ 2x_1 + 4x_2 + x_3 &= m. \end{aligned}$$

3–22. Find the vector X and the maximum value of M that satisfy

$$\begin{aligned} x_i \geq 0 \text{ for } i &= 1, 2, 3, 4 \\ 2x_1 - 4x_2 + x_3 + 3x_4 &= 10 \\ -x_1 + 4x_2 + 2x_3 - x_4 &= 8 \\ 2x_1 + 3x_2 + x_3 + 2x_4 &= M. \end{aligned}$$

3–23. Find nonnegative numbers x_1, x_2, and the maximum M that satisfy

$$\begin{aligned} -3x_1 + 4x_2 &\leq 0 \\ x_1 - 2x_2 &\leq 5 \\ x_1 - x_2 &\leq 6 \\ 2x_1 - x_2 &\leq 14 \\ 5x_1 - x_2 &\leq 39.5 \\ 20x_1 + 7x_2 &\leq 202 \\ 8x_1 + 5x_2 &= M. \end{aligned}$$

Is it always best to choose the pivotal column by the most negative number in the bottom row? Answer by working this problem under both possible choices for the initial pivot. Does the degeneracy in the initial tableau affect the final result? Check your results graphically.

3–24. Run Examples 2 and 3, pages 7–10.

3–25. Maximize $3x_1 + 3x_2$ subject to

$$\begin{aligned} x_1 \geq 0, x_2 &\geq 0 \\ 3x_2 - \quad x_1 &\geq 3 \\ x_2 - \quad 2x_1 &\leq 2. \end{aligned}$$

3–26. Maximize $3x_2 - 2x_1$ subject to

$$\begin{aligned} x_1 \geq 0, x_2 &\geq 0 \\ x_2 - \quad 2x_1 &\geq 2 \\ 3x_2 - \quad x_1 &\leq 3. \end{aligned}$$

3–27. Maximize $4x_2 - 3x_1$ subject to

$$\begin{aligned} x_1 \geq 0, x_2 &\geq 0 \\ -x_1 + \quad 5x_2 &\leq 25 \\ 4x_1 - \quad x_2 &\leq 14 \\ x_1 + \quad 4x_2 &\geq 12. \end{aligned}$$

3–28. Run Problem 1–13, page 14, and find Mrs. Coffman's best buy.

3–29. Maximize $x_1 - x_2 + x_3$ subject to

$$\begin{aligned} x_i \geq 0, i &= 1, 2, 3 \\ -x_1 + \quad x_2 - 2x_3 &\leq 5 \\ 2x_1 - 3x_2 + \quad x_3 &\leq 3 \\ 2x_1 - 5x_2 + 6x_3 &\leq 5. \end{aligned}$$

3–30. Find vector X and the maximum value of M such that

$$\begin{aligned} x_i \geq 0, i &= 1, 2, 3 \\ -x_1 + 4x_2 - 2x_3 &\leq 5 \\ 2x_1 - 3x_2 + \quad x_3 &\leq 3 \\ 2x_1 - 5x_2 + 6x_3 &\leq 5 \\ x_1 + \quad x_2 + \quad x_3 &= M. \end{aligned}$$

3–31. Find vector X and the maximum value of M such that

$$\begin{aligned} x_i \geq 0, i &= 1, 2, 3, 4, 5 \\ -2x_2 + \quad x_3 + x_4 - x_5 &\leq 0 \\ 2x_1 \qquad - 2x_3 + x_4 - x_5 &\leq 0 \\ -x_1 + 2x_2 \qquad + x_4 - x_5 &\leq 0 \\ x_1 + \quad x_2 + \quad x_3 \qquad &= 1 \\ x_1 + \quad x_2 - \quad x_3 + x_4 - x_5 &= M. \end{aligned}$$

3–32. Find both the maximum and the minimum values of $x_1 + x_2 + x_3$ subject to

$$\begin{aligned} x_i \geq 0, i &= 1, 2, 3 \\ 2x_1 - \quad x_2 + 3x_3 &= 9 \\ 5x_1 + 4x_2 - 2x_3 &= 7. \end{aligned}$$

3–33. Find both the maximum and the minimum values of $x_2 + x_3 - x_1$ subject to the same constraints given in Problem 3–32.

3–34. Find nonnegative numbers x_1, x_2 and the minimum value of m

such that

$$\begin{aligned} 7x_1 + 2x_2 &\geq 14 \\ 2x_1 + x_2 &\geq 6 \\ x_1 + x_2 &\geq 5 \\ 3x_1 + 2x_2 &= m. \end{aligned}$$

3–35. Run Example 1, page 3, and check the graphical solution.

3–36. Run Problem 1–15, page 14, and solve the sales problem of Mr. Wise.

3–37. Run Problem 1–16, page 14.

3–38. Find $x_1 \geq 0$ and $x_2 \geq 0$ such that

$$\begin{aligned} 3x_1 + 4x_2 &\leq 12 \\ 2x_1 - 3x_2 &\geq 10 \\ 3x_1 + 5x_2 &= M, \quad \text{a maximum} \end{aligned}$$

3–39. Determine $x_1 \geq 0$ and $x_2 \geq 0$ such that

$$\begin{aligned} 9x_1 - 6x_2 &\geq 3 \\ 8x_1 - 8x_2 &\leq 16 \\ 2x_1 + 2x_2 &= M, \quad \text{a maximum.} \end{aligned}$$

3–40. Find vector X and the minimum value of m subject to

$$\begin{aligned} x_i \geq 0,\ i &= 1, 2, 3 \\ 8x_1 + 2x_2 + x_3 &\geq 3 \\ 3x_1 + 6x_2 + 4x_3 &\geq 4 \\ 4x_1 + x_2 + 5x_3 &\geq 1 \\ x_1 + 5x_2 + 2x_3 &\geq 7 \\ 7x_1 + 3x_2 + 8x_3 &= m. \end{aligned}$$

3–41. Find vector X and the maximum value of M subject to

$$\begin{aligned} x_i \geq 0,\ i &= 1, 2, 3 \\ 3x_1 - 2x_2 + x_3 &\leq 5 \\ 2x_1 + x_2 - x_3 &\leq 3 \\ 9x_1 - 6x_2 + 3x_3 &\leq 15 \\ 12x_1 - 8x_2 + 4x_3 &= M. \end{aligned}$$

3–42. Find the maximum value of M such that

$$\begin{aligned} x_1 \geq 0,\ x_2 &\geq 0 \\ 3x_1 - 2x_2 &\geq 1 \\ x_1 - x_2 &\leq 2 \\ x_1 + x_2 &= M. \end{aligned}$$

3–43. Find vector X and the maximum value of M such that

$$\begin{aligned} x_i \geq 0,\ i &= 1, 2, 3, 4 \\ x_1 + 3x_2 \qquad + x_4 &\leq 4 \\ 2x_1 + x_2 \qquad\qquad &\leq 3 \\ x_2 + 4x_3 + x_4 &\leq 3 \\ 2x_1 + 4x_2 + x_3 + x_4 + 10 &= M. \end{aligned}$$

Hint: The value of M in the initial tableau is not zero.

3–44. Eure's Department Store must reorder 6 popular items. The order will consist of at least a 2 month's supply but not more than a 4 month's supply of the items. In the following table you will find the necessary information where costs and prices are given in dollars and storage space is given in cubic feet per 100 items.

item	costs	selling price	sales per month	storage space
A	1.95	2.29	800	25
B	2.49	2.99	650	30
C	5.85	6.99	300	50
D	2.60	2.99	750	20
E	9.90	12.49	250	72
F	.79	.99	900	10

The current resources of Eure's are $50,000 in cash for investment and 3000 cubic feet of storage space. How many of each of the 6 items should they order to earn the maximum profit neglecting the overhead expenses? Suppose increases are imminent in the wholesale costs. If you were the manager of Eure's and wished to increase the possible profit from this order, would you increase the investment capital by borrowing money or would you add on more storage space?

3–45. Four types of precision rings are turned out in a machine shop on a lathe, a grinder, and a polisher. The following table gives the times in hours per batch required by each ring on each machine. Also listed is the profit in dollars per batch and the maximum number of hours available on each machine per week.

ring	lathe	grinder	polisher	profit
A	8.5	1.5	4.5	90
B	3.0	3.25	8.5	70
C	4.25	4.0	1.75	80
D	2.75	2.75	5.25	60
max. hrs.	48	40	60	

Determine the production schedule and maximum profit for the week. Chop your answers to the tenth of a batch.

3–46. If extra help is available in the machine shop of Problem 3–45 and the grinder can be put to work for an additional 32 hours, then what should the production schedule and profit be? How much of the additional 32 hours is actually used on the grinder?

3–47. Farmer McPherson is considering 5 possible crops for his 100 acres of tillable land. He has $20,000 in credit at the local bank. For this year's growing season Mr. McPherson can count on at most 2000 man-hours of labor and 1200 hours of tractor time. He pays $2.25 per hour for labor plus $3 per hour for the tractor. The following table gives the data in terms of hours per acre or dollars per acre for each crop.

crop	labor time	tractor time	other expenses	profit
barley	16	8	$90	$30
corn	20	16	100	55
beets	48	24	200	100
navy beans	40	24	75	50
potatoes	56	36	200	90

The profit is estimated on an average growing season. The other expenses include the seeds, young plants, fertilizers, insecticides and fungicides. Mr. McPherson allows $50 per acre to plant clover on the acres not used by his crops. What should be planted and how much is the expected profit?

3–48. Suppose the market changes and the expected profit column in Problem 3–47 reads respectively $35, $95, $90, $30, $85. Then what should his planting program and total profit be? In this case is he wasting any of his labor time or tractor time? Has he used all of his credit at the bank?

3–49. Bill Zippo is tuning up his car for the big 500 mile race. He will need one refill of his 22 gallon tank. Bill uses a base gas with an added high octane mix. The mechanic assures him that his engine can stand no more than a ratio of 1 gal. of the high octane mix to 2 gals. of the base gas. By trial Bill discovers that his greatest speed is attained when he uses at least 1 gal. of high octane mix to 3 gals. of base gas. His racer will only go 7 miles per gal. on the base gas but for each gallon of the added high octane mix his milage total is increased by 4/11 miles per gal. The base gas costs 36¢ per gal. and the high octane mix costs 50¢ per gal. How many gallons of each type of gas should Bill purchase for these two tankfuls to attain his maximum speed at the minimum gasoline cost while still finishing the race?

3–50. In Problem 3–49 suppose the high octane mix only increases his milage by $\frac{3}{11}$ miles per gal. Then what would the solution be?

3–51. There are 1000 midshipmen due to graduate from the United States Naval Academy and receive their first duty assignment. The members of the class will each be given one of five possible assignments according to the following conditions. The new officers being assigned to the fleet must be at least one half as many as those that are going to flight school at Pensacola, Florida. There are no more than 250 openings at the nuclear power school and at the graduate school in Monterey, California combined. The number of officers to be assigned to the Marine Corps at Quantico, Virginia must be at least one and a half times the number going to Monterey, California. The number of graduates to be sent to Pensacola is at least equal to the number going to Monterey plus twice the number going to nuclear power school. The sum of the officers going to nuclear power school and to the marines is at least three quarters of the number going to Pensacola. Twice the number assigned to Pensacola, minus one half the number assigned to nuclear power school, minus three times the number assigned to Monterey, plus one half the number assigned to the fleet, minus twice the number assigned to Quantico is at least nine fortieths of the graduating class. Unfortunately not all of the graduates can have their first choice among the duty assignments. Experience has shown that the following numbers are happy with their assignment: 85% of those getting Pensacola, 80% of those getting the nuclear power school, 95% of those getting Monterey, 50% of those getting the fleet, and 85% of those getting the marines. How would you assign the graduating class to satisfy all of the given conditions and create the most happiness?

4

LINEAR PROGRAMMING IN FORTRAN

4.1 Introduction

Most practical problems in linear programming involve many variables leading to large tableaux so that hand computation becomes unbearable as well as too expensive. The algorithms of the simplex method, explained in Chapter 3, are iterative and lead quite naturally to a computer program. A computer program is a set of instructions, in a language understood by the computer, that instructs the computer in the sequence of calculations to be performed.

The most widely used of the many languages available is that of FORTRAN. The FORTRAN language itself has undergone a number of revisions. However, the primary control statements are the same in each version. In this chapter only the subroutines that actually carry out pivoting will be presented. These subroutines can be used in FORTRAN II, FORTRAN IV, or a later version of FORTRAN. The language differences are handled in the main program. The main program controls the input of data to be used, the calls to the necessary subroutines, and the form of the output. It is assumed that the reader has enough familiarity with FORTRAN to write the main program for his machine. The subroutines are usually loaded and stored in the machine memory.

Another popular and rapidly growing method of problem solving is by *time sharing* on a group of large computers coordinated so as to handle many users at the same time. These users contact the computer center from remote terminals that may be hundreds of miles away. A remote terminal is a teletype machine that is connected through special telephone lines to the computer. A user types his data and instructions into the computer and receives his response directly on his own teletype.

The computer center usually involves several computers working together. One computer handles the incoming calls from many users and the sequencing of information directed to the central processor. The central processor or main computer does the actual computations involved. The results are stored, perhaps a small bit at a time, until the communications computer is ready to give a response to a particular user. In this way dozens of programs can be processed simultaneously. The central processor works continuously doing a small portion of each user's computations at a time as sequenced in and out by the communications computer. The responses are so rapid that each individual user feels as if he had the whole computer to himself. The tremendous advantage to the user of time sharing is found in his conversational access to a high power computer. This interaction allows the user to make small changes in a program, such as varying a parameter, and receive immediate results.

Time sharing is very convenient for many types of linear programs. This is especially true for integer programming and parametric programming that are discussed in Chapters 8 and 9. Most time sharing computer centers offer the FORTRAN language. Another popular time sharing language is BASIC. It is a simple matter to convert the following FORTRAN subroutines into BASIC or any other language that is familiar and available. A glossary of FORTRAN commands explaining their use is included in the back of this book (pages 212–230) for the novice in FORTRAN programming.

4.2 Pivoting the Extended Tableau in FORTRAN

Following the notation of Chapter 3, variable Y (I, J) will stand for the tableau entry in the *i*th row and *j*th column. The size of the tableau will be indicated by integer variables M, the number of rows, and N, the number of columns. Our pivot position is the *p*th row and *q*th column. In order to make these integer variables, they will be designated by IP and IQ respectively. If IP, IQ, M, N, and tableau Y are known, then the following subroutine will carry out one pivoting iteration.

```
      SUBROUTINE PIVEX (IP, IQ, M, N, Y)
      DIMENSION Y (25, 25)
      A=Y (IP, IQ)
      DO 10 J=1, N
   10 Y (IP, J)=Y (IP, J)/A
      DO 50 I=1, M
      IF (I-IP) 30, 50, 30
   30 B=Y (I,IQ)
      DO 40 J=1, N
   40 Y (I, J)=Y(I,J)-B*Y (IP, J)
   50 CONTINUE
      RETURN
      END
```

The purpose of SUBROUTINE PIVEX is to produce a unit column vector in the pivotal column. The brackets above mark the DO loops so that the logical sequence can easily be followed. Statement 10 leaves a one in the pivot position when J=IQ. Statement 40 produces zeros in the pivotal column except at the pivotal row. The IF statement was used to skip the pivotal row in the outer loop. The inner loop carries out the Gauss-Jordan elimination across the ith row while the outer loop merely advances the computation from row to row. Thus we are left with a unit column vector in the pivotal column.

PIVEX can be used in a number of applications. For example, if the pivoting is required to proceed down the main diagonal, it will solve a consistent and independent set of n linear equations in n unknowns. Care must be taken to arrange the equations so that a zero will not appear on the main diagonal at a pivot position during any iteration. This may be done whenever the equations are both independent and consistent. The original tableau should consist of the n by $(n + 1)$ matrix formed from the coefficients of the unknowns plus an additional column on the right for the constant terms. After pivoting down the main diagonal the solution vector appears in the right-hand column, and the remainder of the tableau is the identity matrix.

Another application of PIVEX is found in the inversion of a nonsingular matrix. The inverse of a matrix is only defined for nonsingular square arrays. By definition the **inverse** of an $N \times N$ matrix A is another $N \times N$ matrix called A^{-1} such that $AA^{-1} = I$, where I is the $N \times N$ identity matrix. Again there must be no zeros on the main diagonal at a pivot position. The original tableau should consist of the given matrix on the left and the corresponding identity matrix on the right. For example, if the matrix is

$$\begin{bmatrix} Y_{11} & Y_{12} \\ Y_{21} & Y_{22} \end{bmatrix},$$

then the initial tableau is

$$\begin{bmatrix} Y_{11} & Y_{12} & 1 & 0 \\ Y_{21} & Y_{22} & 0 & 1 \end{bmatrix}.$$

By pivoting down the main diagonal the original matrix is reduced to the identity, and the identity matrix is transformed into the inverse. For those familiar with matrix theory this statement can be proved by noting that the steps in pivoting are all elementary matrix operations. That is, each step can be performed by multiplication on the left by an appropriate *elementary matrix*. Let E be the product of these elementary matrices and let A be the given matrix.

Then,

$$EA = I$$
$$EAA^{-1} = IA^{-1}$$
$$EI = A^{-1}$$

Thus the set of *elementary operations* that transforms A into I, the identity, simultaneously transforms I into A^{-1}.

Of course PIVEX can also be used to pivot the extended tableau of a linear program to produce a new basic feasible solution. However, we will solve most linear programs by pivoting the condensed tableau.

4.3 Pivoting the Condensed Tableau in FORTRAN

In order to pivot the condensed tableau we introduce two new subscripted variables, K(J), J=1 to N−1, and L(I), I=1 to M−1. The K(J) stand for the subscripts of the variables out of the basis that appear across the top of our tableau. The L (I) stand for the subscripts of the variables that are in the basis and appear to the left of our tableau. The initial values of K (J) and L (I) are read in by the main program along with the initial tableau. After the pivot position, IP, IQ, has been determined, the following subroutine will carry out the calculations.

```
      SUBROUTINE PIVCO (Y, K, L, M, N, IP, IQ)
      DIMENSION Y (30, 30), K (30), L (30)
      R=1./Y (IP, IQ)
      DO 10 J=1, N
   10 Y (IP, J)=R*Y (IP, J)
      Y (IP, IQ)=R
      DO 50 I=1, M
      IF (I-IP) 20, 50, 20
   20 DO 40 J=1, N
      IF (J-IQ) 30, 40, 30
   30 Y (I, J)=Y (I, J)-Y (I, IQ)*Y (IP, J)
   40 CONTINUE
      Y (I, IQ)= -R*Y (I, IQ)
   50 CONTINUE
      X=K (IQ)
      K (IQ)=L (IP)
      L (IP)=X
      RETURN
      END
```

The five steps of the Condensed Tableau Pivoting Algorithm from Chapter 3 are evident in the subroutine. The first statement after the dimension declaration stores the reciprocal of the pivot in location R for future use. Statement 10 in the first loop corresponds to step 2 of our algorithm. The next statement after 10 resets the pivot and corresponds to step 1. Statement 30 carries out step 4 after the two IF statements skip the pivotal row and skip the pivotal column respectively. The inner loop computes the entries across the *i*th row. The outer loop moves the computation from row to row through the tableau. The statement between 40 and 50 corresponds to step 3. It is in the outer loop and computes the entry in the pivotal column for each row except the pivotal row. Finally step 5 is carried out by the last three statements of the subroutine before returning to the main program. Note that the integer K (IQ) is temporarily stored in location X so that it will not be destroyed when replaced with integer L (IP).

4.4 Automatic Pivoting Subroutines

We wish to have the computer find the pivot position and automatically continue to pivot until the final tableau is reached. The main program need only print out the final tableau at the conclusion of a problem. Three additional subroutines are used to make the whole process automatic.

The first of these is our column search which determines a pivotal column if any is available. If there is no possible pivotal column, then the following subroutine returns IQ= −1 to the main program.

```
      SUBROUTINE COL (Y, K, M, N, IQ)
      DIMENSION Y (30, 30), K (30)
      IQ= -1
      B=0.
      NN=N-1
      DO 30 J=1, NN
      IF (K (J)) 30, 10, 10
   10 IF (Y (M, J)) 20, 30, 30
   20 IF (Y (M, J)-B) 25, 30, 30
   25 B=Y (M, J)
      IQ=J
   30 CONTINUE
      RETURN
      END
```

Notice that the first IF statement prevents any artificial variable from being brought back into the basis. If K (J) is negative that column is skipped and never used again in choosing a pivot. Statement 10 picks out the negative numbers of the last row. Statements 20 and 25 assure us that the negative number of greatest absolute value is chosen before leaving IQ equal to the number of that column. Thus the value of IQ returned to the main program either signifies the pivotal column or the fact that none is available. If IQ remains negative then the main program should print the current values of Y (I, J) as the final tableau.

The next subroutine is our row search that picks out a pivotal row if one is available. If not, ROW returns IP= −1 to the main program.

```
      SUBROUTINE ROW (Y, M, N, IP, IQ)
      DIMENSION Y (30, 30)
      IP= -1
      S= .1 E19
      MM=M-1
      DO 80 I=1, MM
      IF (Y (I, IQ)) 80, 80, 10
   10 R=Y (I, N)/Y (I, IQ)
      IF (S-R) 80, 80, 20
   20 S=R
      IP=I
   80 CONTINUE
      RETURN
      END
```

The first IF statement skips any negative or zero entries in the pivotal column. Thus only positive θ ratios are considered in statement 10. The second IF statement assures us that the smallest of the θ ratios is chosen before setting IP equal to that row. If there is a tie for least our program picks out the first of the tying rows encountered. If ROW and COL return positive values to the main program then the next command in the main program should be

```
CALL PIVCO (Y, K, L, M, N, IP, IQ)
```

to carry out the pivoting. At any time that either ROW or COL return a negative value, the final tableau has been reached and should be printed out.

One more subroutine is needed to distinguish between phase one pivoting and phase two pivoting. The main program must know if any artifical variables remain in the basis. If so then it must call the column search at row M. This is phase one where negative numbers are looked for in the last row. As soon as the last artificial variable leaves the basis, the main program must call the column search at row M−1. This is phase two where negative numbers are sought in the first of the two objective rows. Of course if the original problem had no artificial variables then only one objective row is used. In order to allow for these three possible branches, SUBROUTINE LOOK is called before each column search until the artificial variables are removed.

```
      SUBROUTINE LOOK (L, M, V)
      DIMENSION L (30)
      DO 10 I=1, M
      IF (L (I)) 30, 10, 10
   10 CONTINUE
      V=1.
      RETURN
   30 V= -1.
      RETURN
      END
```

For convenience all of the original rows of the tableau are numbered on the left so that L (I) is defined from 1 to M. The objective rows may be given any positive number. SUBROUTINE LOOK returns V=−1 if any L (I) is negative, and it returns V=1 if all L (I) are positive. At any stage the main program knows if there are artificial variables in the basis by a call to LOOK and testing the value of V.

This completes the subroutines necessary to carry out an automatic pivoting program. Care must be taken in the main program to use the row search only over the first M − 2 rows for problems with a double objective row. We never pivot on one of the objective rows. Upon reaching the final tableau the program should print out the K (J) and L (I) in their respective positions along with the Y (I, J) for a complete interpretation of the results. Finally, it is nice to put in a counter to tell the number of times that the original tableau was pivoted.

Problems Chapter 4

4–1. Write a FORTRAN program that will pivot a given tableau down its main diagonal to produce an $M \times M$ identity matrix.
Hint: The principal step is a DO loop that will set the pivot and call the SUBROUTINE PIVEX.

```
      DO 60 K=1, M
      IP=K
      IQ=K
      CALL PIVEX (IP, IQ, M, N, Y)
   60 CONTINUE
```

Use your FORTRAN *program to solve the following six systems of linear equations by pivoting down the main diagonal.*

4–2. $3x_1 + 2x_2 - x_3 = -4$
$2x_1 - 5x_2 - 4x_3 = 0$
$8x_1 - 3x_2 - 2x_3 = 8$

4–3. $3x_1 - 4x_2 + 2x_3 = 8$
$x_1 + 3x_2 - 5x_3 = 13$
$6x_1 - 4x_2 - 7x_3 = 9$

4–4. $$\begin{aligned} 5x_1 - x_2 + x_3 &= 12 \\ x_1 + 2x_2 + 3x_3 &= 15 \\ 4x_1 + 2x_2 - 3x_3 &= 5 \end{aligned}$$

4–5. $$\begin{aligned} x_1 - 2x_2 - 3x_3 + 4x_4 &= 5 \\ -6x_1 + 7x_2 + 8x_3 + 9x_4 &= 12 \\ 9x_1 - 8x_2 + 7x_3 - 6x_4 &= 16 \\ 5x_1 - 4x_2 - 3x_3 - 2x_4 &= 10 \end{aligned}$$

4–6. $$\begin{aligned} 5x_1 - x_2 + x_3 &= 11 \\ x_1 + 2x_2 + 3x_3 &= 17 \\ 4x_1 + 2x_2 - 3x_3 &= 3 \end{aligned}$$

4–7. $$\begin{aligned} 2x_1 - 4x_2 + 3x_3 - 5x_4 + 2x_5 &= -13 \\ 3x_1 + 2x_2 + 5x_3 + 4x_4 + 6x_5 &= -7 \\ -4x_1 + 5x_2 - 6x_3 + 3x_4 - 2x_5 &= 15 \\ 5x_1 + 3x_2 - 4x_3 - 8x_4 + 7x_5 &= 9 \\ -6x_1 - x_2 + 9x_3 - 3x_4 - 5x_5 &= -12 \end{aligned}$$

4–8. Write a FORTRAN program that will invert a given nonsingular matrix without zeros at a pivot position.

4–9. The nth order Hilbert matrix may be defined as follows.

$$\begin{bmatrix} 1 & \frac{1}{2} & \frac{1}{3} & \cdots & \frac{1}{n} \\ \frac{1}{2} & \frac{1}{3} & \frac{1}{4} & \cdots & \frac{1}{(n+1)} \\ \frac{1}{3} & \frac{1}{4} & \frac{1}{5} & \cdots & \frac{1}{(n+2)} \\ \vdots & & & & \vdots \\ \frac{1}{n} & \frac{1}{(n+1)} & \frac{1}{(n+2)} & \cdots & \frac{1}{(2n-1)} \end{bmatrix}$$

Invert the second order Hilbert matrix by hand to be clear on the process used in section 4.2. As a check multiply the Hilbert matrix by its inverse to see if the product is the appropriate identity matrix.

4–10. Invert the third order Hilbert matrix by using your FORTRAN program.
Hint: The matrix should be constructed by a double DO loop in your program. The only data is the order N. Such a set of nested DO loops follows.

```
         B= -1.
         DO 80 I=1, N
         B=B+1
         DO 80 J=1, N
         A=J
   80    Y(I, J)=1./(A+B)
```

4–11. Invert the fourth order Hilbert matrix with your program.

4–12. Invert the fifth order Hilbert matrix. Notice the size of the numbers in your inverse. The determinant of the Hilbert matrix is small so the numbers in its inverse grow very rapidly with the order. Round-off error soon becomes important. You may check on the size of the error by having your machine multiply the Hilbert matrix by its inverse and seeing how close the product is to the identity matrix.

4–13. There is a serious flaw in the program of Problem 4–1, namely, that it blows up if the pivot happens to be zero or close to zero. Correct this flaw by the following technique. Search down the pivotal column below the pivot for the entry of greatest absolute value. If this absolute value is greater than that of the pivot, then interchange the pivotal row with this new row before carrying out the pivoting. Thus the largest numerical value available will always be used for the pivot. In particular zero will be avoided if possible. If the pivot and all entries below it are zero, then the matrix is singular and the system of equations has no unique solution.

Hint: After setting the pivot position, the following loop will determine which row has the greatest absolute value in the pivotal column.

```
         X=ABS (Y (IP, IQ))
         L=IP
         NN=IP+1
         DO 60 MM=NN, M
         IF (ABS (Y (MM, IQ)) -X) 60, 60, 55
   55    X=ABS (Y (MM, IQ))
         L=MM
   60    CONTINUE
```

After this loop is executed L will denote the desired row. If X is not too small and L is different from IP, then interchange row IP with row L. Finally call PIVEX and then continue to the next pivot position.

4–14. Use your revised program from Problem 4–13 to solve the following system.

$$9x_2 - 2x_3 = 14$$
$$4x_1 + 5x_2 + 3x_3 = -1$$
$$6x_1 + 7x_2 + 4x_3 = 2$$

4–15. Does the following system have a unique solution?

$$x_1 + 2x_2 + 5x_3 = 21$$
$$2x_1 + 4x_2 + 3x_3 = 7$$
$$3x_1 + 6x_2 + 2x_3 = -2$$

4–16. Solve:

$$4x_2 - 7x_3 = 13$$
$$3x_1 + 5x_3 = 0$$
$$6x_1 + 11x_2 = 8.$$

4–17. Solve:

$$x_1 + \tfrac{1}{2}x_2 + \tfrac{1}{3}x_3 = 32$$
$$\tfrac{1}{2}x_1 + \tfrac{1}{3}x_2 + \tfrac{1}{4}x_3 = 22$$
$$\tfrac{1}{3}x_1 + \tfrac{1}{4}x_2 + \tfrac{1}{5}x_3 = 17.$$

4–18. Solve:

$$x_1 + \tfrac{1}{2}x_2 + \tfrac{1}{3}x_3 + \tfrac{1}{4}x_4 = 167$$
$$\tfrac{1}{2}x_1 + \tfrac{1}{3}x_2 + \tfrac{1}{4}x_3 + \tfrac{1}{5}x_4 = 125$$
$$\tfrac{1}{3}x_1 + \tfrac{1}{4}x_2 + \tfrac{1}{5}x_3 + \tfrac{1}{6}x_4 = 101$$
$$\tfrac{1}{4}x_1 + \tfrac{1}{5}x_2 + \tfrac{1}{6}x_3 + \tfrac{1}{7}x_4 = 85.$$

4–19. Write a FORTRAN program that will pivot a given condensed tableau at a given position. Hint: The principal step is an input command that will allow the operator to type in the pivot position, (IP, IQ). Then a call to SUBROUTINE PIVCO completes the job.

4–20. Write the main FORTRAN program that will automatically pivot a given tableau until the final tableau is reached. The pivot position is found by a call to COL and if successful followed by a call to ROW. If artificial variables are present then SUBROUTINE LOOK is used for branching. Provide for three branches. One branch is for tableaux with a single objective row. The other two branches are for phase one and phase two pivoting of tableaux with a double objective row.

5

MATRIX GAMES AND THE CONCEPT OF DUALITY

5.1 Introduction and Definition of a Matrix Game

Since duality arises so naturally in the setting of a matrix game, we will look into a little elementary game theory. The discussion will be restricted to games with a finite number of outcomes that can be solved as linear programs on a computer. The development of a theory of game strategy began in the 1920's and continues to this day. The theory made its greatest advance under John von Neumann, who deserves credit for the extensive development as presented in his classic work *Theory of Games and Economic Behavior* published jointly with Oskar Morganstern.[1] In 1928 von Neumann gave the first proof of the central theorem of matrix games, the famous Minimax Theorem. At the time linear programming was not available, so his proof used the fixed point theorems of L. E. J. Brouwer, and other advanced mathematics.

The connection between game theory and linear programming is not obvious, but was recognized by George Dantzig who offered an elementary constructive proof of the Minimax Theorem in 1956.[2] We will demonstrate the connection between the two when we express the solution to a matrix game in a linear programming format.

Definition 1. A **matrix game** is a two-person game described by an $m \times n$ pay-off matrix. One person A will choose rows while the other person B chooses columns. An entry, Y_{ij}, in the pay-off matrix tells the amount of money to be paid to player A by player B whenever the ith row and jth column are chosen independently by the two players.

[1] John von Neumann and Oskar Morganstern, *Theory of Games and Economic Behavior*, 1953.
[2] George B. Dantzig, "Constructive Proof of the Min-Max Theorem," *Pacific Journal of Mathematics*, 6 (1) (1956): 25.

The entries Y_{ij} in the pay-off matrix may be positive, negative, or zero. A positive pay-off represents a gain for player A while a negative pay-off is a loss for A, that is, a pay-off to player B. A zero entry is a tie or no pay-off.

A matrix game as described here is an example of a finite, zero-sum, two-person game. There are only a finite number of choices for each player. The term **zero-sum** means that the algebraic sum of the money won and lost by all players is zero. Wealth is neither created nor destroyed but merely redistributed in a zero-sum game. A slot machine is not a zero-sum game because the house takes a healthy cut before any pay-off to the players.

We are interested in the plays that will win the most money for player A, and also the plays that are best for player B. Of course, B is looking for negative pay-offs in the matrix. He wishes to pay the least amount possible to A by picking the most negative pay-offs. The picture of duality arises in solving for the best possible strategies for each of the players.

5.2 Analysis of 2 × 2 Matrix Games

A simple example of a matrix game is the matching of pennies between two persons.

Example 1. The following pay-off matrix shows that player A receives a penny when the coins match and he pays out a penny when the coins differ.

		B heads	B tails
A	heads	1	−1
	tails	−1	1

The problem for both A and B is how often to play heads and how often to play tails. This leads us to the idea of relative frequency or probability.

Definition 2. The **probability** of making a particular play i is a rational number x, $0 \leq x \leq 1$, that is the ratio of the number of times i will be played to the total number of plays of the game.

We assume A and B can choose their probability of playing heads at will. Thus in 100 matches if A decides to play heads 63 times in some unknown order, then his probability of heads on a particular match is $x = .63$. Since on his remaining matches A must play tails, A's probability of tails on a particular match is $1 - x = .37$. Such a probability could be achieved by placing a spinner on a circle divided into 100 sectors. An arbitrary 63 sectors should be marked heads and the remaining 37 sectors marked tails. Then a spin of the needle constitutes a play for A. A probability of zero for some choice means that choice will never be played, and a probability of one means that choice will be taken every time. The problem is to determine a pair of probabilities that is optimal for A and another pair that is optimal for B.

Definition 3. A **mixed strategy** for n choices is an ordered n-tuple of probabilities, $(x_1, x_2, \ldots, x_n)$, such that $x_1 + x_2 + \cdots + x_n = 1$.

Definition 4. The ***i*th pure strategy** is a mixed strategy with 1 in the ith place and 0 elsewhere.

In the case of the penny match let A's mixed strategy be $(x_1, x_2) = (x, 1 - x)$ and B's mixed strategy be $(y, 1 - y)$, where $0 \leq x \leq 1$ and $0 \leq y \leq 1$.

		B	
		y	$1 - y$
A	x	1	-1
	$1 - x$	-1	1

Let z be A's expectation per play, that is, the amount A can expect to gain on the average per play of the game. With probability x, A can expect to gain 1¢ under B's probability y while losing 1¢ with probability $(1 - y)$. Thus, from the first row, A can expect to gain

$$x[1y + (-1)(1 - y)].$$

Likewise with probability $(1 - x)$, A can expect to lose 1¢ with probability y and gain 1¢ with probability $(1 - y)$. A's expected gain in the second row is then

$$(1 - x)[(-1)y + 1(1 - y)].$$

The sum of these two, z, is the total expected gain of A per play.

$$z = x[1y + (-1)(1-y)] + (1-x)[(-1)y + 1(1-y)]$$
$$= x(2y-1) + (1-x)(1-2y)$$
$$\langle 1 \rangle \qquad = 4xy - 2x - 2y + 1$$
$$= 4(xy - \tfrac{1}{2}x - \tfrac{1}{2}y) + 1$$
$$\langle 2 \rangle \qquad = 4(x - \tfrac{1}{2})(y - \tfrac{1}{2}).$$

In the general case of an $m \times n$ pay-off matrix we have the following definition.

Definition 5. The **mathematical expectation** of player A is the double summation,

$$z = \sum_{i=1}^{m} \sum_{j=1}^{n} x_i Y_{ij} y_j,$$

where $(x_1, x_2, \ldots, x_m)$ and $(y_1, y_2, \ldots, y_n)$ are any mixed strategies for A and B, and Y_{ij} are the entries in the pay-off matrix.

An analogous expression with the summations reversed can be written down for the mathematical expectation of player B. However, each sum has only a finite number of terms so the order is immaterial. This means that the definition of mathematical expectation of B is identical to that of A.

The object of the game from A's viewpoint is to make his expectation z as large as possible. A negative value of z indicates a loss for A or a gain by player B. B's objective is to minimize z so that he pays the least amount possible to A. Of course, if B can make z negative then he wins that amount on the average per play.

To see what strategy is optimal consider in equation $\langle 2 \rangle$ that player A chooses his probability of heads to be $x < \frac{1}{2}$. Then the factor $(x - \frac{1}{2})$ is negative. Player B after noticing this trend can always win by making his factor $(y - \frac{1}{2})$ positive, that is, by choosing $y > \frac{1}{2}$. On the other hand if A chooses $x > \frac{1}{2}$ then $(x - \frac{1}{2})$ is positive. Player B can again always win by making $(y - \frac{1}{2})$ negative, choosing in this case $y < \frac{1}{2}$. The only way that player A can avoid losing is by choosing $x = \frac{1}{2}$. A's optimal strategy is then $(\frac{1}{2}, \frac{1}{2})$. By exactly the same reasoning, we arrive at B's optimal strategy to be $(\frac{1}{2}, \frac{1}{2})$. Thus, the best play for both A and B is to take an independent, random toss to assure that they play heads or tails with equal frequency. ●

The optimal mixed strategies are computed from a conservative view. That is, player A asks the question, "What is my minimum return under a given strategy?" He then tries to maximize this minimum. By choosing $x = \frac{1}{2}$ in the penny match game, A guarantees his expectation to be at least zero no matter what B does. Thus on the average A can do no worse than break even. Player B asks the question, "What is my maximum pay-off to A under a given strategy?" He then tries to minimize this maximum. The value B arrives at in this way is the most he will have to pay out no matter what A does.

Definition 6. The **value** of a matrix game **to player** A is his expectation computed under his optimal strategy no matter what B does. The game **value to player** B is B's expectation computed under his optimal strategy no matter what A does.

A's game value is found by determining the strategy that will maximize his minimum expectation, and B's game value is found by determining the strategy that will minimize his maximum expectation.

Definition 7. The game is called **fair to player** A if his value is zero, and the game is **fair to player** B if his value is zero.

The value of the penny match game is $z = 0$ for both players and so the game is fair to both. We will see that this is not a mere coincidence.

Example 2. Consider a different pay-off matrix:

		B	
		y	$1-y$
A	x	-3	6
	$1-x$	2	-5

If A and B choose strategy $(\frac{1}{2}, \frac{1}{2})$ then $z = 0$ and the game seems to be fair. Let us compute $z = f(x, y)$ and see if the game is fair.

$$\begin{aligned}
z &= x[-3y + (1-y)6] + (1-x)[2y - 5(1-y)] \\
&= x(6 - 9y) + (1-x)(7y - 5) \\
\langle 3 \rangle \qquad &= -16xy + 11x + 7y - 5 \\
&= -16(xy - \tfrac{11}{16}x - \tfrac{7}{16}y) - 5 \\
&= -16(x - \tfrac{7}{16})(y - \tfrac{11}{16}) + \tfrac{77}{16} - 5 \\
\langle 4 \rangle \qquad &= -16(x - \tfrac{7}{16})(y - \tfrac{11}{16}) - \tfrac{3}{16}
\end{aligned}$$

Player A controls factor $(x - \frac{7}{16})$ and B controls factor $(y - \frac{11}{16})$. If either one makes the value of their factor different from zero the other player can take the advantage. Thus the optimal strategies are $(\frac{7}{16}, \frac{9}{16})$

for A and $(\frac{11}{16}, \frac{5}{16})$ for B. The value of the game is $-\frac{3}{16}$ to both players which is not fair to player A. On the average player A will lose 3¢ every 16 plays. ●

5.3 Computation of Strategies[3]

The computation may be simplified by using calculus. Consider the locus of $z = f(x, y)$ as a surface in three dimensional space. The surfaces represented by equations ⟨**1**⟩ and ⟨**3**⟩ are hyperbolic paraboloids. The hyperbolic paraboloid is distinguished by its *saddle* shape and a point called the saddle point. The saddle point is the minimum of parabolas on the surface opening upward and the maximum of other parabolas on the surface opening downward. Cross sections orthogonal to the planes of these parabolas are hyperbolas. Figure 5–1 represents the hyperbolic paraboloid $z = xy$.

Relative extreme points and saddle points are found on differentiable surfaces by the following sequence of theorems. The subscripts stand for partial derivatives with respect to the variables in the subscript.

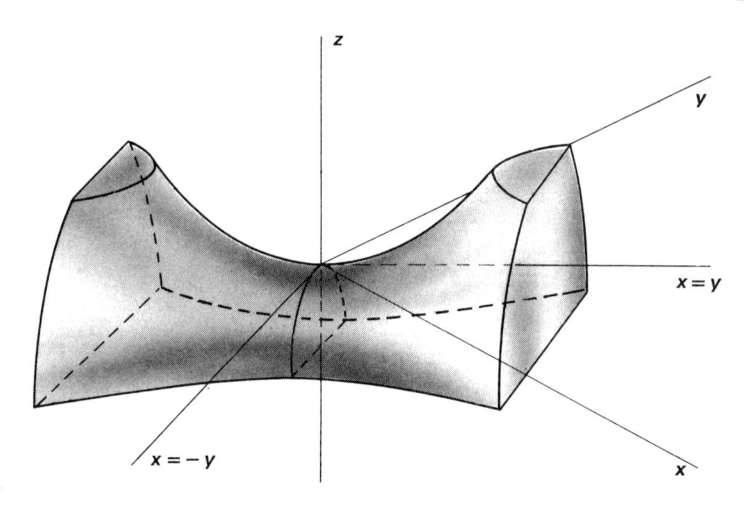

FIGURE 5–1

Theorem 5–1. *If $f(x, y)$ has continuous first partial derivatives, necessary conditions for f to have an extreme at (a, b) are $f_x = 0$ and $f_y = 0$ at (a, b).*

A set of sufficient conditions for a relative extreme is given by the following theorem.

[3] This section may be eliminated by those not familiar with two variable calculus.

Theorem 5–2. *If $f(x, y)$ has continuous second partial derivatives, if $f_x = f_y = 0$ at (a, b), and if $f^2{}_{xy} - f_{xx}f_{yy} < 0$ at (a, b), then the point (a, b) is an extreme of f. The extreme will be a maximum if $f_{xx} < 0$ and a minimum if $f_{xx} > 0$ at (a, b).*

Theorem 5–3.[4] *If $f(x, y)$ has continuous second partial derivatives, if $f_x = f_y = 0$ at (a, b), and if $f^2{}_{xy} - f_{xx}f_{yy} > 0$ at (a, b), then the point (a, b) is a saddle point of f.*

Applying Theorem 5–3 to Example 1, page 71, we have

$$
\begin{aligned}
&f(x, y) = 4xy - 2x - 2y + 1 \\
&f_x = 4y - 2 = 0, \qquad f_y = 4x - 2 = 0 \\
&y = \tfrac{1}{2}, \qquad\qquad\quad x = \tfrac{1}{2} \\
&f_{xx} = 0, \qquad f_{xy} = 4, \qquad f_{yy} = 0 \\
&f^2{}_{xy} - f_{xx}f_{yy} = 16 > 0.
\end{aligned}
$$

Thus the point $(\frac{1}{2}, \frac{1}{2}, 0)$ on the surface is a saddle point. From Example 2, page 74, we find

$$
\begin{aligned}
&f(x, y) = -16xy + 11x + 7y - 5 \\
&f_x = -16y + 11 = 0, \qquad f_y = -16x + 7 = 0 \\
&y = \tfrac{11}{16}, \qquad\qquad\qquad x = \tfrac{7}{16} \\
&f_{xx} = 0, \qquad f_{xy} = -16, \qquad f_{yy} = 0 \\
&f^2{}_{xy} - f_{xx}f_{yy} = 256 > 0.
\end{aligned}
$$

This checks that the point $(\frac{7}{16}, \frac{11}{16}, -\frac{3}{16})$ is a saddle point. In both cases the coordinates of the saddle point give the optimal strategies and the value of the game. This will be true in general provided the saddle point occurs over the unit square, $0 \leq x \leq 1$, $0 \leq y \leq 1$.

5.4 An Example of Pure Strategies

Let us look at one more example of a 2×2 matrix game.

Example 3.

		B	
		y	$1 - y$
A	x	1	2
	$1 - x$	-2	1

[4] For proof see D. V. Widder, *Advanced Calculus*, second edition, p. 128.

For this game A's expectation is

$$\begin{aligned} z &= x[y + 2(1 - y)] + (1 - x)[-2y + (1 - y)] \\ &= x(2 - y) + (1 - x)(1 - 3y) \\ &= 2xy + x - 3y + 1 \\ &= 2(xy + x/2 - 3y/2) + 1 \\ &= 2(x - \tfrac{3}{2})(y + \tfrac{1}{2}) + \tfrac{3}{2} + 1 \\ \langle 5 \rangle \qquad &= 2(x - \tfrac{3}{2})(y + \tfrac{1}{2}) + \tfrac{5}{2} \end{aligned}$$

The situation is quite different. While A still controls the first factor and B controls the second factor, neither one can make their factor equal to zero. B notices that A's factor, $(x - \frac{3}{2})$, is always negative no matter what probability A picks. Therefore, B will make his factor, $(y + \frac{1}{2})$, as large as possible to take advantage of the negative product. B will choose $y = 1$. Of course, A wishes to make his negative factor as small as possible. The best he can do is to pick $x = 1$. Thus, the optimal strategies are $(1, 0)$ and both players will always play heads. ●

Definition 8. A game is called **strictly determined** if the optimal strategies of both players are pure.

Example 3 had optimal pure strategies and is a strictly determined game. From equation $\langle 5 \rangle$ we see that the value of the game is $z = 1$ for both players. Player A will win at least 1¢ on each play of the game. A might win more than 1¢ if B were to use any strategy other than $(1, 0)$. If B uses $(1, 0)$ then he is guaranteed to pay out no more than 1¢ to A. The saddle surface

$$z = 2xy + x - 3y + 1$$

has a saddle point at $(\frac{3}{2}, -\frac{1}{2}, \frac{5}{2})$, but in this case its coordinates are outside of the range of probabilities.

The same solution can be reached directly from the pay-off matrix by the following reasoning. The poorest A can do in the first row is to win 1¢ while the worst he can do in the second row is to lose 2¢. He will pick the first row to get the maximum of his minimum possible gains. Player B looks at columns. The most B can lose in the first column is 1¢ while his maximum loss in the second column is 2¢. He will pick the first column to get the minimum of his maximum possible losses. One cent is both the maximum of A's minimums and the minimum of B's maximums. This common value is known as a **minimax** value. A minimax is the common value that both players will pay or receive under optimal pure strategies. Thus, if a matrix has a minimax value, it is the value of the game and the game is strictly determined. While most matrices do not have a minimax entry, we will show that the value of all games is the same for both A and B. That is, the maximum that A will win on the average per play is the same as the minimum B will lose on the average per play when each uses his best strategy. This result is the Minimax Theorem proved by John von Neumann.

Problems Section 5.4

For the following matrix games, find the game value and the optimal strategies by first finding the saddle surface $z = f(x, y)$. Also state whether the game is fair. In case the game is strictly determined, give the location of the minimax entry.

5–1. (a) $\begin{bmatrix} 5 & -4 \\ -3 & 2 \end{bmatrix}$ (b) $\begin{bmatrix} -4 & 3 \\ 6 & 0 \end{bmatrix}$

5–2. (a) $\begin{bmatrix} 0 & 2 \\ -2 & 0 \end{bmatrix}$ (b) $\begin{bmatrix} -5 & 4 \\ 3 & -1 \end{bmatrix}$

5–3. (a) $\begin{bmatrix} 1 & 0 \\ 2 & -3 \end{bmatrix}$ (b) $\begin{bmatrix} 3 & -2 \\ -2 & 3 \end{bmatrix}$

5–4. (a) $\begin{bmatrix} -5 & 6 \\ 5 & -6 \end{bmatrix}$ (b) $\begin{bmatrix} 1 & 2 \\ 3 & 4 \end{bmatrix}$

5–5. (a) $\begin{bmatrix} -3 & 4 \\ 2 & -3 \end{bmatrix}$ (b) $\begin{bmatrix} 2 & -3 \\ -3 & 4 \end{bmatrix}$

(c) What effect does interchanging the rows have upon the results?
(d) Would the corresponding effect be true for interchanging columns?

5–6. (a) $\begin{bmatrix} a & 0 \\ 0 & d \end{bmatrix}$ (b) $\begin{bmatrix} 0 & b \\ c & 0 \end{bmatrix}$

5–7. Given the pay-off matrix

$$\begin{bmatrix} a & b \\ c & d \end{bmatrix}$$

(a) Find the value of the game, z, and the optimal strategies. Assuming that the game is not strictly determined, these quantities may be expressed in terms of the numbers a, b, c, and d.
(b) How is the value z affected by adding the same constant k to each member of the pay-off matrix?
(c) How is the value z affected by multiplying each member of the pay-off matrix by the same constant k?
(d) Are the optimal strategies affected by these two operations in parts (b) and (c)? The important conclusion here is true in general for any size matrix and will be used in Section 5.6.

5.5 A Graphical Solution of $m \times 2$ Matrix Games

In the special case of $m \times 2$ matrix games A's optimal strategy may be found by graphing his strategy polygon. A similar dual analysis will find B's optimal strategy in the case of $2 \times n$ matrix games. Consider the following example.

Example 4.

$$A \overset{B}{\begin{bmatrix} 3 & 1 \\ 1 & -2 \\ 2 & 6 \\ -4 & 2 \end{bmatrix}}$$

Let A's strategy be (t_1, t_2, t_3, t_4) where $t_i \geq 0$ for $i = 1, 2, 3, 4$ and $t_1 + t_2 + t_3 + t_4 = 1$. First consider the pay-off to A in the extreme cases of B's two possible pure strategies. Let x equal A's pay-off against B's pure strategy (1, 0). Then

$$\langle \mathbf{1} \rangle \qquad x = 3t_1 + 1t_2 + 2t_3 - 4t_4 .$$

Let y equal A's pay-off against B's pure strategy (0, 1). Then

$$\langle \mathbf{2} \rangle \qquad y = 1t_1 - 2t_2 + 6t_3 + 2t_4$$

If B uses a mixed strategy $(s, 1 - s)$ then his pay-off to A, called z, will be the convex combination of pay-offs x and y,

$$z = sx + (1 - s)y.$$

It is easy to see that the pay-off z is always between x and y. For example if $x \leq y$, then

$$x = sx + (1 - s)x \leq z \leq sy + (1 - s)y = y.$$

The inequalities are just reversed in case $x \geq y$. This proves the following theorem.

Theorem 5–4. *B's minimum pay-off to A is the smaller of x or y.*

We wish to graph the set of points (x, y) representing all possible pay-offs to player A under B's pure strategies. The vertices of this set are found by substituting the four possible pure strategies of A into equations $\langle \mathbf{1} \rangle$ and $\langle \mathbf{2} \rangle$. These points correspond respectively to the rows of the pay-off matrix. The entire set, $\{(x, y)\}$, satisfying the restrictions on t_i, is the smallest convex set containing these four pure strategy points of A. It is the **convex hull** of the pure strategy points.

The set may be shown graphically by drawing the smallest convex polygon, R, containing the pure strategy points. The resulting polygon known as A's **strategy polygon** is shown in Figure 5–2.

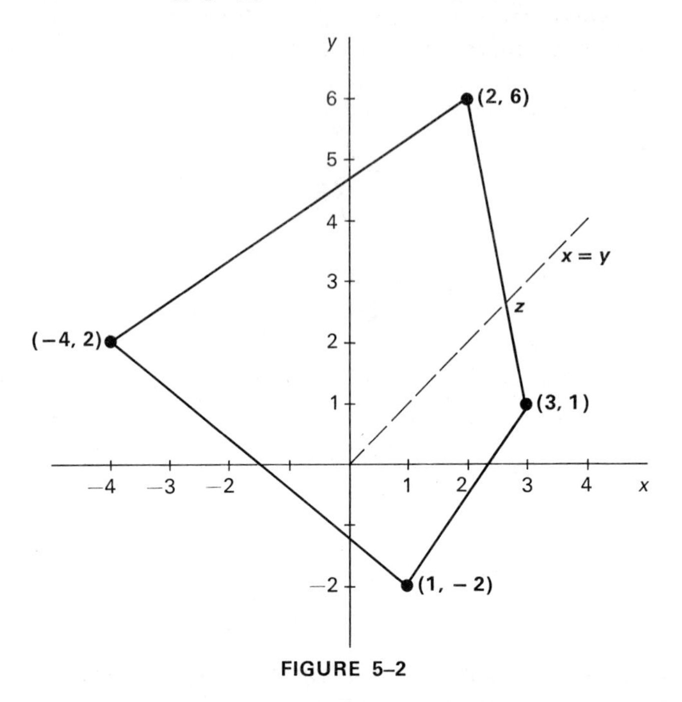

FIGURE 5–2

By Theorem 5–4, for any point of A's strategy polygon, B will minimize his pay-off to A by picking the smaller of its coordinates x or y. The best choice for A is to pick a strategy point (x, y) whose smaller coordinate is a maximum over R. This may be done by proceeding out from the origin along the line $x = y$ until it intersects the polygon boundary at a point with the largest abscissa x. In our example that boundary has a negative slope and the desired point is the intersection z shown. Points above line $x = y$ in the polygon have a smaller abscissa and points below $x = y$ in the polygon have a smaller ordinate than z. If the boundary in question has a positive slope then we may continue out along this boundary until we reach the vertex with the largest abscissa. In that case the game would be strictly determined.

In our case z represents the optimum pay-off, and either one of its coordinates gives the value of the game for A. The parametric equations of the boundary from (3, 1) to (2, 6) are

$$x = 3t_1 + 2(1 - t_1)$$
$$y = 1t_1 + 6(1 - t_1)$$
$$t_2 = 0,\ t_4 = 0,\ t_3 = 1 - t_1.$$

Probabilities t_2 and t_4 are 0 since the boundary line of R through z is a convex combination of vertices (3, 1) and (2, 6). In the parametric equations note that probability $t_1 = 1$ corresponds to vertex (3, 1) and $t_1 = 0$ or $t_3 = 1$ corresponds to vertex (2, 6).

Setting x equal to y and solving for t_1, we find $t_1 = \frac{2}{3}$, $t_3 = (1 - t_1) = \frac{1}{3}$. Thus A's optimal strategy is $(\frac{2}{3}, 0, \frac{1}{3}, 0)$. The game value for A is $x = y = \frac{8}{3}$. By the Minimax Theorem $\frac{8}{3}$ is also the game value for player B. ●

Since the game value will always occur at a boundary point of A's strategy polygon, we have the following theorem.

***Theorem* 5–5.** *In $m \times 2$ matrix games player A needs at most a convex combination of two of his m pure strategies. Similarly, in $2 \times n$ matrix games, B needs at most a convex combination of two of his n pure strategies.*

An alternate graph to the strategy polygon is the line graph in the following example.

Example 5. Consider the following $2 \times n$ game.

$$\begin{array}{cc} & B \\ A & \begin{bmatrix} 5 & -2 & 2 & -1 \\ -2 & 4 & -1 & 1 \end{bmatrix} \end{array}$$

Let A's strategy be $(1 - t, t)$. Let x_j, $j = 1, 2, 3, 4$, be the pay-offs to A under B's four possible pure strategies. Then A wishes to maximize the minimum of the linear functions

$$x_j = Y_{1j}(1 - t) + Y_{2j}t$$

for $j = 1, 2, 3, 4$ where Y_{ij} are the entries in the pay-off matrix. These linear functions of t are respectively

$$\begin{aligned} x_1 &= 5(1 - t) - 2t \\ x_2 &= -2(1 - t) + 4t \\ x_3 &= 2(1 - t) - t \\ x_4 &= -(1 - t) + t. \end{aligned}$$

The four lines are graphed in Figure 5–3 by noting that each line passes through $(0, Y_{1j})$ and $(1, Y_{2j})$. The minimum function of these lines is the sequence of line segments in bold in Figure 5–3. The maximum of this minimum function occurs at the intersection of lines x_3 and x_4.

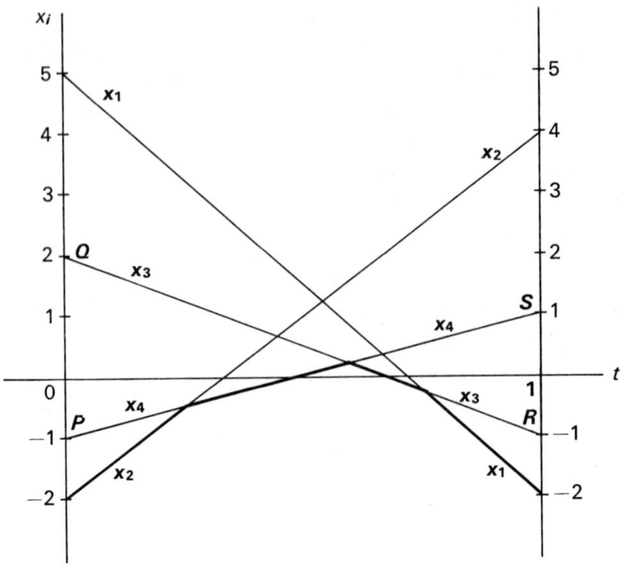

FIGURE 5–3

Setting $x_3 = x_4$ gives

$$2(1-t) - t = -(1-t) + t$$
$$2 - 3t = -1 + 2t$$
$$t = \tfrac{3}{5}.$$

Thus A's optimal strategy is $(\frac{2}{5}, \frac{3}{5})$ and the value of the game is $x_3 = x_4 = \frac{1}{5}$.

B's optimal strategy may be found by finding the ratio in which height $\frac{1}{5}$ divides the line segment PQ from x_4 to x_3 at $t = 0$. This ratio is

$$(\tfrac{6}{5})/(\tfrac{9}{5}) = \tfrac{2}{3}.$$

The strategy for B corresponding to the ratio of 2 to 3 is $(0, 0, \frac{2}{5}, \frac{3}{5})$. The same strategy could be found by using the ratio in which height $\frac{1}{5}$ divides the line segment RS from x_3 to x_4 at $t = 1$. This ratio is

$$(\tfrac{6}{5})/(\tfrac{4}{5}) = \tfrac{3}{2}.$$

The ratio is reversed since the order of the corresponding lines is reversed at $t = 1$. ●

In the case of $m \times 2$ matrix games the corresponding lines would be drawn for A's m pure strategies. Then the game value would be the minimum height of the maximum of these linear functions.

Problems Section 5.5

For the following matrix games find the game value and player A's optimal strategy from A's strategy polygon.

5–8. $\begin{bmatrix} 4 & 1 \\ -1 & 3 \end{bmatrix}$

5–9. (a) $\begin{bmatrix} -1 & 1 \\ -3 & 7 \\ 3 & -2 \\ 1 & 6 \end{bmatrix}$ (b) $\begin{bmatrix} -1 & 1 \\ -3 & 7 \\ 2 & 5 \\ 1 & -1 \end{bmatrix}$

5–10. (a) $\begin{bmatrix} 2 & 4 \\ 5 & 1 \\ 3 & 6 \end{bmatrix}$ (b) $\begin{bmatrix} -5 & 1 \\ 3 & -2 \\ 4 & 7 \end{bmatrix}$

5–11. (a) $\begin{bmatrix} -8 & 5 \\ 2 & -4 \\ 6 & -1 \end{bmatrix}$ (b) $\begin{bmatrix} 4 & 1 \\ 3 & -2 \\ 2 & -4 \end{bmatrix}$

5–12. $\begin{bmatrix} 1 & 6 \\ -2 & 3 \\ -1 & 0 \\ 2 & -4 \\ 7 & 3 \end{bmatrix}$

For the next set of matrix games, find the game value and player B's optimal strategy from B's strategy polygon. In this case choose the point of B's polygon for which the larger coordinate is minimized.

5–13. $\begin{bmatrix} 4 & 1 \\ -1 & 3 \end{bmatrix}$

5–14. (a) $\begin{bmatrix} 0 & 3 & 4 \\ 3 & 2 & 0 \end{bmatrix}$ (b) $\begin{bmatrix} 2 & -1 & 5 \\ 5 & -2 & -1 \end{bmatrix}$

5–15. (a) $\begin{bmatrix} -2 & 3 & -1 \\ 4 & 3 & 1 \end{bmatrix}$ (b) $\begin{bmatrix} 0 & -4 & 4 \\ 3 & 1 & -3 \end{bmatrix}$

5–16. $\begin{bmatrix} -8 & 1 & 4 & 6 \\ 6 & 7 & -3 & 2 \end{bmatrix}$

5–17. $\begin{bmatrix} -4 & 2 & 2 & -2 & 5 \\ 1 & 5 & -1 & -3 & -2 \end{bmatrix}$

5–18. Solve Problem 5–16 by drawing the appropriate line graph.

5–19. Solve Problem 5–17 by drawing its line graph.

5–20. Solve Problem 5–12 by drawing its line graph.

5.6 Linear Programming for Matrix Games and the Definition of Duality

In this section we will develop a linear program to solve the general $m \times n$ matrix game. In the process we will discover the underlying principle of duality. The Fundamental Duality Theorem will then imply von Neumann's Minimax Theorem as a corollary.

Example 6. Let us solve the following matrix game.

$$\begin{array}{cc} & B \\ A & \begin{bmatrix} 1 & -2 & 1 \\ -1 & 3 & -2 \\ -1 & -2 & 3 \end{bmatrix} \end{array}$$

There are two problems. A's problem is to find his optimal strategy and his maximum pay-off under optimal strategies. B's problem is to determine his optimal strategy and his minimum pay-off to A under optimal strategies. Let (x_1, x_2, x_3) be B's strategy and let (x_4, x_5, x_6) be A's strategy where $x_i \geq 0$, $i = 1, 2, \ldots, 6$, and

$$\langle 1 \rangle \qquad x_1 + x_2 + x_3 = 1$$

$$\langle 2 \rangle \qquad x_4 + x_5 + x_6 = 1.$$

Because it will be easier to work with a positive pay-off, we add 2 to each entry in the pay-off matrix. This will remove all negative entries and insure that the pay-off to A is a positive number. Recall, from the problems of Section 5.4, that the addition of a constant to the pay-off matrix does not change the optimal strategies. However, the game value will be 2 more than the value of the original game. Thus we will solve the following game and then subtract 2 from its game value.

$$\begin{array}{cc} & B \\ A & \begin{bmatrix} 3 & 0 & 3 \\ 1 & 5 & 0 \\ 1 & 0 & 5 \end{bmatrix} \end{array}$$

We attack B's problem first. Let L_1, L_2, L_3 be respectively B's pay-offs to A under A's three pure strategies, $(1, 0, 0)$, $(0, 1, 0)$, and $(0, 0, 1)$. Then

$$\langle 3 \rangle \qquad \begin{aligned} L_1 &= 3x_1 + 0x_2 + 3x_3 \\ L_2 &= 1x_1 + 5x_2 + 0x_3 \\ L_3 &= 1x_1 + 0x_2 + 5x_3. \end{aligned}$$

Let L be the largest of the three pay-offs, L_1, L_2, and L_3. Since A will choose L, B wishes to minimize L. By replacing the L_i in equations ⟨**3**⟩ with L, we have type I linear constraints

$$\langle \mathbf{4} \rangle \qquad \begin{aligned} 3x_1 + 3x_3 &\le L \\ 1x_1 + 5x_2 &\le L \\ 1x_1 + 5x_3 &\le L. \end{aligned}$$

Consider each of the constraints in ⟨**4**⟩ divided by L. The pay-offs to A were forced to be positive so that L would be greater than zero. Thus the sense of the inequalities is preserved. Now rename the variables as follows:

$$x_1' = x_1/L, \qquad x_2' = x_2/L, \qquad x_3' = x_3/L.$$

Also divide equation ⟨**1**⟩ through by L and then make the given change of variables. In order for B to minimize L, he will maximize $1/L$. B's problem may now be stated as a linear program.

Find nonnegative numbers x_1', x_2', x_3', subject to the constraints

$$\langle \mathbf{5} \rangle \qquad \begin{aligned} 3x_1' + 3x_3' &\le 1 \\ 1x_1' + 5x_2' &\le 1 \\ 1x_1' + 5x_3' &\le 1, \end{aligned}$$

such that the number

$$x_1' + x_2' + x_3' = 1/L = M$$

is a maximum.

Let the three slack variables be x_4', x_5', and x_6'. Then B's problem leads directly to the condensed tableau

	1	2	3	
4	3	0	3	1
5	1	5	0	1
6	1	0	5	1
	−1	−1	−1	0

Note that the tableau is make up of the pay-off matrix, with a column of ones on the right, and a row of minus ones at the bottom along with zero as the initial value of M. Three pivots bring us to the optimal tableau. The first pivot is on position (1, 1) giving

	4	2	3	
1	$\frac{1}{3}$	0	1	$\frac{1}{3}$
5	$-\frac{1}{3}$	5	-1	$\frac{2}{3}$
6	$-\frac{1}{3}$	0	4	$\frac{2}{3}$
	$\frac{1}{3}$	-1	0	$\frac{1}{3}$

Next, pivot on (2, 2).

	4	5	3	
1	$\frac{1}{3}$	0	1	$\frac{1}{3}$
2	$-\frac{1}{15}$	$\frac{1}{5}$	$-\frac{1}{5}$	$\frac{2}{15}$
6	$-\frac{1}{3}$	0	4	$\frac{2}{3}$
	$\frac{4}{15}$	$\frac{1}{5}$	$-\frac{1}{5}$	$\frac{7}{15}$

Finally, pivot on (3, 3) for the following optimal tableau.

	4	5	6	
1	$\frac{5}{12}$	0	$-\frac{1}{4}$	$\frac{1}{6}$
2	$-\frac{1}{12}$	$\frac{1}{5}$	$\frac{1}{20}$	$\frac{1}{6}$
3	$-\frac{1}{12}$	0	$\frac{1}{4}$	$\frac{1}{6}$
	$\frac{1}{4}$	$\frac{1}{5}$	$\frac{1}{20}$	$\frac{1}{2}$

From the optimal tableau $M = \frac{1}{2}$ and $L = 1/M = 2$. From our transformation equations

$$x_1 = Lx_1' = 2(\tfrac{1}{6}) = \tfrac{1}{3}$$
$$x_2 = Lx_2' = 2(\tfrac{1}{6}) = \tfrac{1}{3}$$
$$x_3 = Lx_3' = 2(\tfrac{1}{6}) = \tfrac{1}{3}.$$

Thus B's optimal strategy is $(\frac{1}{3}, \frac{1}{3}, \frac{1}{3})$. Remembering that 2 was added to the original pay-off matrix, we find the game value from B's viewpoint to be $L - 2 = 0$. Zero is the minimum B may expect to pay A on the average per play of the game.

Next let us solve A's problem. Let S_1, S_2, S_3 be respectively A's pay-offs under B's three pure strategies, (1, 0, 0), (0, 1, 0,), and (0, 0, 1). Then

$$\langle 6 \rangle \qquad \begin{aligned} S_1 &= 3x_4 + 1x_5 + 1x_6 \\ S_2 &= 0x_4 + 5x_5 + 0x_6 \\ S_3 &= 3x_4 + 0x_5 + 5x_6 . \end{aligned}$$

Let S be the smallest of the three pay-offs, S_1, S_2, and S_3. Since B will choose S, A wishes to maximize S. By replacing the S_i in equations $\langle 6 \rangle$ with S, we have type II inequalities.

$$\langle 7 \rangle \qquad \begin{aligned} 3x_4 + 1x_5 + 1x_6 &\geq S \\ 5x_5 \qquad &\geq S \\ 3x_4 \qquad + 5x_6 &\geq S. \end{aligned}$$

Consider each of the constraints in $\langle 7 \rangle$ divided by S. Once again the sense of the inequalities is preserved because all pay-offs were forced to be positive and $S > 0$. Rename the variables as follows,

$$x_4' = x_4/S, \qquad x_5' = x_5/S, \qquad x_6' = x_6/S.$$

Also divide equation $\langle 2 \rangle$ through by S and make the change of variables. In order for A to maximize S, he will minimize $1/S$. A's problem may now be stated as a linear program.

Find nonnegative numbers x_4', x_5', x_6', subject to the constraints

$$\langle 8 \rangle \qquad \begin{aligned} 3x_4' + 1x_5' + 1x_6' &\geq 1 \\ 5x_5' \qquad &\geq 1 \\ 3x_4' \qquad + 5x_6' &\geq 1, \end{aligned}$$

and such that the number

$$x_4' + x_5' + x_6' = 1/S = m$$

is a minimum.

Let the three slack variables be x_1', x_2', x_3', and let the three artificial variables be x_{-7}, x_{-8}, x_{-9}. Change the minimizing problem to a maximizing problem by setting $m = -M$. Let the artificial maximum be

$$\overline{M} = M - N(x_{-7} + x_{-8} + x_{-9})$$

or

$$\overline{M} + x_4' + x_5' + x_6' + N(x_{-7} + x_{-8} + x_{-9}) = 0.$$

N is an arbitrarily large positive number so that the artificial variables will leave the basis and then $\overline{M}$ will equal M. This leads to the following initial condensed tableau for A.

	4	5	6	1	2	3	
−7	3	1	1	−1	0	0	1
−8	0	5	0	0	−1	0	1
−9	3	0	5	0	0	−1	1
$\overline{M}$	1	1	1	0	0	0	0
	−6	−6	−6	1	1	1	−3

We choose to break the ties by pivoting on (1, 1) with the following result.

	−7	5	6	1	2	3	
4	$\frac{1}{3}$	$\frac{1}{3}$	$\frac{1}{3}$	$-\frac{1}{3}$	0	0	$\frac{1}{3}$
−8	0	5	0	0	−1	0	1
−9	−1	−1	4	1	0	−1	0
$\overline{M}$	$-\frac{1}{3}$	$\frac{2}{3}$	$\frac{2}{3}$	$\frac{1}{3}$	0	0	$-\frac{1}{3}$
	2	−4	−4	−1	1	1	−1

Next, pivot on (2, 2) and drop the first column.

	−8	6	1	2	3	
4	$-\frac{1}{15}$	$\frac{1}{3}$	$-\frac{1}{3}$	$\frac{1}{15}$	0	$\frac{4}{15}$
5	$\frac{1}{5}$	0	0	$-\frac{1}{5}$	0	$\frac{1}{5}$
−9	$\frac{1}{5}$	4	1	$-\frac{1}{5}$	−1	$\frac{1}{5}$
$\overline{M}$	$-\frac{2}{15}$	$\frac{2}{3}$	$\frac{1}{3}$	$\frac{2}{15}$	0	$-\frac{7}{15}$
	$\frac{4}{5}$	−4	−1	$\frac{1}{5}$	1	$-\frac{1}{5}$

Again drop the first column and then pivot on the first entry of the third row. This pivot corresponds to position (3, 3) of the original matrix and produces the final tableau.

	-9	1	2	3	
4	$-\frac{1}{12}$	$-\frac{5}{12}$	$\frac{1}{12}$	$\frac{1}{12}$	$\frac{1}{4}$
5	0	0	$-\frac{1}{5}$	0	$\frac{1}{5}$
6	$\frac{1}{4}$	$\frac{1}{4}$	$-\frac{1}{20}$	$-\frac{1}{4}$	$\frac{1}{20}$
$\overline{M} = M$	$-\frac{1}{6}$	$\frac{1}{6}$	$\frac{1}{6}$	$\frac{1}{6}$	$-\frac{1}{2}$
	1	0	0	0	0

The artificial variables have now been removed from the basis and may be ignored along with the bottom row. In particular $\overline{M} = M$. Rewrite the optimal tableau accordingly.

	1	2	3	
4	$-\frac{5}{12}$	$\frac{1}{12}$	$\frac{1}{12}$	$\frac{1}{4}$
5	0	$-\frac{1}{5}$	0	$\frac{1}{5}$
6	$\frac{1}{4}$	$-\frac{1}{20}$	$-\frac{1}{4}$	$\frac{1}{20}$
M	$\frac{1}{6}$	$\frac{1}{6}$	$\frac{1}{6}$	$-\frac{1}{2}$

The solution to A's problem may now be read off. $M = -\frac{1}{2}$ so the minimum, $m = -M = \frac{1}{2}$. Then $S = 1/m = 2$.

Transforming back to the original variables, we have

$$x_4 = Sx_4' = 2(\tfrac{1}{4}) = \tfrac{1}{2}$$
$$x_5 = S_5' = 2(\tfrac{1}{5}) = \tfrac{2}{5}$$
$$x_6 = Sx_6' = 2(\tfrac{1}{20}) = \tfrac{1}{10}.$$

Thus A's optimal strategy is $(\frac{1}{2}, \frac{2}{5}, \frac{1}{10})$. As before we must subtract 2 from the game value to get back to the original game. A's game value is $S - 2 = 0$. Zero is the maximum A may expect to win on the average per play of the game. The game is fair to both players. As is predictable by the Minimax Theorem, A's maximum gain is the same as B's minimum loss under optimal strategies. ●

Far more important than the solution to the game is the comparison between A's final tableau and B's final tableau. Carefully compare the two.

B's optimal tableau

	4	5	6	
1	$\frac{5}{12}$	0	$-\frac{1}{4}$	$\frac{1}{6}$
2	$-\frac{1}{12}$	$\frac{1}{5}$	$\frac{1}{20}$	$\frac{1}{6}$
3	$-\frac{1}{12}$	0	$\frac{1}{4}$	$\frac{1}{6}$
	$\frac{1}{4}$	$\frac{1}{5}$	$\frac{1}{20}$	$\frac{1}{2}$

A's optimal tableau

	1	2	3	
4	$-\frac{5}{12}$	$\frac{1}{12}$	$\frac{1}{12}$	$\frac{1}{4}$
5	0	$-\frac{1}{5}$	0	$\frac{1}{5}$
6	$\frac{1}{4}$	$-\frac{1}{20}$	$-\frac{1}{4}$	$\frac{1}{20}$
	$\frac{1}{6}$	$\frac{1}{6}$	$\frac{1}{6}$	$-\frac{1}{2}$

It was not by accident that we labeled the slack variables of one problem the same as the main variables of the other problem. The fact that the slack and main variables are interchanged is a part of the duality picture. It should be noticed that the values of all six primed variables can be read off, directly opposite their subscripts, from either one of the two optimal tableaux. In other words the solution to either one of the problems automatically gives the solution to both. The matrix in the upper left hand corner of one tableau is the **negative transpose** of the other. That is, any row of one matrix is the negative of the corresponding column of the other. Finally the optimum value in the lower right hand corner of one tableau is the negative of the optimum value in the other tableau. It is time to define dual tableaux.

Definition 9. The **dual** of an $m \times n$ **tableau** is an $n \times m$ tableau satisfying the following four conditions:

1. The basic and the nonbasic variables are interchanged. This interchanges the subscripts $L(I)$ with $K(J)$.
2. The matrix in the upper left hand corner is the negative transpose of the corresponding matrix.
3. The bottom row and right-hand column are interchanged.
4. The value in the lower right-hand corner is the negative of its corresponding value.

The definition of dual tableaux is clearly symmetric and involutory. *Symmetric* means that if T_1 is dual to T_2 then T_2 is dual to T_1. *Involutory* means that the dual of the dual is the original tableau.

Recall that the process of pivoting was derived from Gaussian elimination. Thus, we are able to preserve equivalent systems of equations from tableau to tableau. Therefore, the set of linear equations represented in our final tableau is equivalent to the original set of equations in the initial tableau. Now if two problems have their final tableaux the dual of one another, we might suspect that the original tableaux were dual. In order to compare the initial tableau of player A's problem with that of B, we first avoid artificial variables by multiplying the constraints ⟨**8**⟩ by minus one. This leads to the following statement of A's problem.

Example 7. Find nonnegative numbers x_4', x_5', x_6', subject to the constraints

$$\langle \mathbf{9} \rangle \qquad \begin{aligned} -3x_4' - 1x_5' - 1x_6' &\leq -1 \\ -5x_5' \qquad &\leq -1 \\ -3x_4' \qquad - 5x_6' &\leq -1, \end{aligned}$$

and such that $x_4' + x_5' + x_6' = m = -\overline{M}$, where m is a minimum and $\overline{M}$ is a maximum.

Using constraints ⟨**5**⟩ and ⟨**9**⟩ the two initial tableaux appear as

	1	2	3	
4	3	0	3	1
5	1	5	0	1
6	1	0	5	1
M	−1	−1	−1	0

B's initial tableau

	4	5	6	
1	−3	−1	−1	−1
2	0	−5	0	−1
3	−3	0	−5	−1
$\overline{M}$	1	1	1	0

A's initial tableau

Indeed the two initial tableaux are dual and we will define the two problems to be dual linear programs. ●

Definition 10. Two linear **programs** are **dual** provided: the slack variables of one are the main variables of the other; type I and type II inequalities are interchanged; the entire coefficient matrix, including the objective function, of one is the negative transpose of the other; and finally the objective maximization is interchanged with minimization.

As an example of Definition 10, the following two programs are dual.

Find $x_1 \geq 0$, $x_2 \geq 0$, $x_3 \geq 0$ satisfying.

$$\begin{aligned} a_{11}x_1 + a_{12}x_2 + a_{13}x_3 &\leq b_1 \\ a_{21}x_1 + a_{22}x_2 + a_{23}x_3 &\leq b_2, \end{aligned}$$

and

$$c_1x_1 + c_2x_2 + c_3x_3 = M, \qquad \text{a maximum.}$$

Find $x_4 \geq 0$, $x_5 \geq 0$ satisfying

$$\begin{aligned} a_{11}x_4 + a_{21}x_5 &\geq c_1 \\ a_{12}x_4 + a_{22}x_5 &\geq c_2 \\ a_{13}x_4 + a_{23}x_5 &\geq c_3, \end{aligned}$$

and

$$b_1x_4 + b_2x_5 = m, \qquad \text{a minimum.}$$

If the initial tableaux of the above problems are written down without artificial variables, they will satisfy the duality of Definition 9. Assuming that all of the constraints of the maximization problem are type I inequalities, the two Definitions 9 and 10 are equivalent.

5.7 The Dual Simplex Method

In trying to solve player A's problem without artificial variables we run into a difficulty. A's initial tableau is already optimal but not feasible. The initial value of the basic variables x_1', x_2', x_3' is -1. We wish to transform this tableau by a method, known as the **Dual Simplex Method**, that will gain feasibility while preserving the optimality. The Dual Simplex Method utilizes pivoting, but the pivot will be chosen by the following algorithm.

The Dual Simplex Algorithm

1. Starting with a condensed tableau that is optimal but not feasible, choose the pivotal row by picking the most negative value in the last column.
2. Using only the negative entries in that pivotal row, compute the θ ratios. They are the ratios of the entries in the last row to the corresponding entries in the pivotal row. All of these ratios are negative.
3. Determine the pivotal column by picking the algebraically largest ratio of step (2). This will be the ratio whose absolute value is the smallest.
4. Carry out a pivoting iteration by the condensed tableau algorithm of Chapter 3.
5. Repeat the first four steps until the tableau is feasible or until no new pivot can be found.

In order to distinguish a program from its dual, we agree to call the maximization problem the *primal program* and call its dual the *dual program*. The dual program is then a minimization problem. In the case of our matrix game with players A and B, the primal program is B's problem since his initial tableau corresponds to a maximization, while A's problem is the dual program since his initial tableau corresponds to a minimization. The pivot chosen by the Dual Simplex Algorithm in the dual program is the *dual* of the corresponding pivot chosen by the Simplex Algorithm in the primal program. This means that if the pivot is at (i, j) in the primal tableau then it will occur at (j, i) in the dual tableau. If the initial tableau of player B is pivoted by the Simplex Method, and if the dual initial tableau of player A is pivoted by the Dual Simplex Method, then each pair of tableaux will be dual through the final pair. This important idea should be checked out on the tableaux of A and B in the game problem of the last section. We will prove in the next chapter that dual pivots of a pair of dual tableaux lead to a pair of dual tableaux. This is the heart of the Fundamental Duality Theorem.

Problems Section 5.6–5.7

Solve the following matrix games by pivoting the condensed tableaux. State the optimal strategies of both players and the value of the game.

5–21. (a) $\begin{bmatrix} 3 & 5 \\ 2 & 1 \\ 5 & 2 \end{bmatrix}$ (b) $\begin{bmatrix} 5 & -2 \\ 4 & 6 \\ -3 & 4 \end{bmatrix}$

5–22. (a) $\begin{bmatrix} 0 & 3 & 5 \\ 1 & 0 & 3 \end{bmatrix}$ (b) $\begin{bmatrix} -1 & 5 & 4 \\ 3 & 6 & -1 \end{bmatrix}$

5–23. (a) $\begin{bmatrix} 1 & 2 \\ 3 & 4 \\ 5 & 3 \end{bmatrix}$ (b) $\begin{bmatrix} 2 & -2 & -4 \\ -3 & 2 & 3 \end{bmatrix}$

5–24. (a) $\begin{bmatrix} 2 & 1 \\ -2 & 6 \\ 3 & 5 \end{bmatrix}$ (b) $\begin{bmatrix} 3 & 4 & 6 \\ 1 & 5 & 4 \end{bmatrix}$

5–25. (a) $\begin{bmatrix} 5 & 3 & 2 \\ 3 & 4 & 1 \end{bmatrix}$ (b) $\begin{bmatrix} 4 & 2 \\ 1 & 3 \\ 3 & 4 \end{bmatrix}$

(c) Check the solutions to (a) and (b) by solving them graphically.

5–26. Solve the following matrix game for optimal strategies by setting up A's problem and then pivoting according to the Dual Simplex Algorithm.

$$\begin{bmatrix} 1 & -2 & 2 \\ -3 & 4 & -1 \\ 5 & -2 & -1 \end{bmatrix}$$

5–27. Carry out the same directions in Problem 5–26 on the following game.

$$\begin{bmatrix} 1 & -1 & 0 \\ 0 & 1 & -1 \\ -1 & 0 & 1 \\ 1 & 0 & -1 \\ -1 & 1 & 1 \end{bmatrix}$$

5–28. Solve the following game for optimal strategies and game value.

$$\begin{bmatrix} 1 & 1 & -1 & 0 & 1 \\ 1 & 0 & 0 & 1 & -1 \\ -1 & 0 & 1 & -1 & 0 \\ 0 & 1 & -1 & 0 & 1 \\ 1 & -1 & 0 & 1 & 1 \end{bmatrix}$$

5–29. Solve the following skew-symmetric[5] matrix game for optimal strategies and game value.

$$\begin{bmatrix} 0 & 1 & -2 & 3 & -4 \\ -1 & 0 & 3 & -4 & 5 \\ 2 & -3 & 0 & 5 & -6 \\ -3 & 4 & -5 & 0 & 7 \\ 4 & -5 & 6 & -7 & 0 \end{bmatrix}$$

Hint: Add 10 to each pay-off entry.

5–30. From Problem 5–29 what might you conjecture about the game value and the optimal strategies in the case of a skew-symmetric pay-off matrix? Can you prove your conjecture?

5–31. A popular childrens' game is *Stone, Scissors, and Paper.* Each player independently picks one of the three items. If there is a tie then there is no pay-off. Otherwise stone "breaks" scissors and stone is paid one, paper "covers" stone and paper is paid two, while scissors "cuts" paper and scissors is paid three. Set up a 3×3 pay-off matrix for this game. Solve the game for the optimal strategies of both players. Is the game fair?

5–32. Two players independently choose a number from one to three. The pay-off is computed by adding the two numbers chosen. If the sum is even, player B pays player A that amount, and if the sum is odd, then A pays B that amount. Set up the pay-off matrix and solve the game for optimal strategies. Is this game fair?

5–33. Two players independently choose a number from one to four. The pay-off is computed by adding the two numbers chosen. If the sum is even, player B pays player A that amount, and if the sum is odd, then A pays B that amount. Set up the pay-off matrix and

[5] A square matrix is skew-symmetric if $Y_{ij} = -Y_{ji}$ for all i and j.

solve the game for optimal strategies. Is this a fair game? Are the optimal strategies unique?

5–34. Two players independently choose a number from one to four. The pay-off is computed by taking the absolute value of the difference of the two numbers chosen. If this absolute value is even, player B pays player A that amount, and if this absolute value is odd, then A pays B that amount. Set up the pay-off matrix for this game. Solve the game for its optimal strategies and game value.

5–35. Two players independently choose a number from one to four. If there is a tie between the numbers chosen then player B pays player A that tying amount. If the numbers chosen are different then the absolute value of this difference is computed. When this absolute value is even, player B pays player A that amount, and when this absolute value is odd, A pays B that amount. Set up the pay-off matrix for this game. Solve the game for its optimal strategies and game value.

6

DUALITY THEOREMS

6.1 The Fundamental Duality Theorem

In Chapter 5 we found that the solution of a matrix game for the strategies of each player led to a pair of dual linear programs. Furthermore, we noticed that the solution to both programs could be found from the solution to either one of the two dual programs. We will now prove this basic idea in general for any pair of dual linear programs.

Theorem 6–1. (Fundamental Duality Theorem) *For a pair of dual linear programs, if one solution exists, the solutions to both programs may be found by solving either one of the two programs.*

Proof. The first job is to show that dual pivots transform dual programs into dual programs. Consider the following pair of dual condensed tableaux.

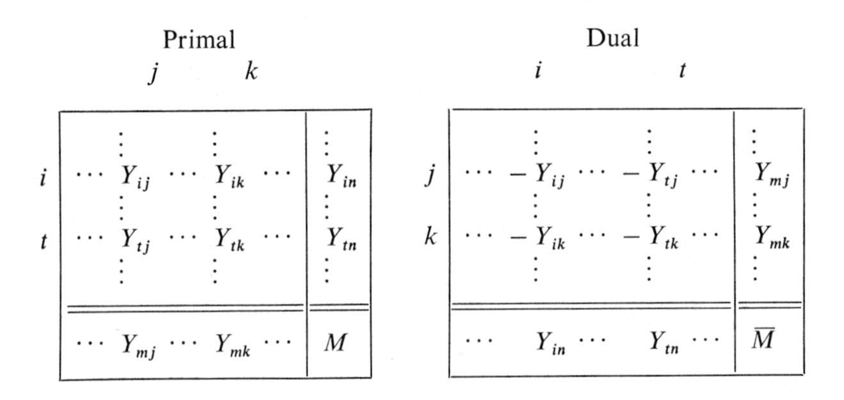

Pivoting the primal tableau at position (i, j) gives

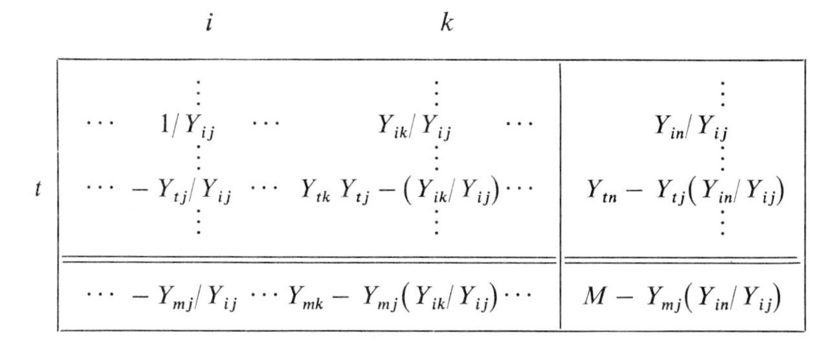

	i	k	
	$\cdots \quad 1/Y_{ij} \quad \cdots$	$Y_{ik}/Y_{ij} \quad \cdots$	Y_{in}/Y_{ij}
t	$\cdots \quad -Y_{tj}/Y_{ij} \quad \cdots$	$Y_{tk}\, Y_{tj} - (Y_{ik}/Y_{ij}) \cdots$	$Y_{tn} - Y_{tj}(Y_{in}/Y_{ij})$
	$\cdots \quad -Y_{mj}/Y_{ij} \quad \cdots$	$Y_{mk} - Y_{mj}(Y_{ik}/Y_{ij}) \cdots$	$M - Y_{mj}(Y_{in}/Y_{ij})$

Pivoting the dual tableau at the dual position (j, i) gives

	j	t	
i	$\cdots \quad -1/Y_{ij} \quad \cdots$	$Y_{tj}/Y_{ij} \quad \cdots$	$-Y_{mj}/Y_{ij}$
k	$\cdots \quad -Y_{ik}/Y_{ij} \quad \cdots$	$-Y_{tk} + Y_{ik}(Y_{tj}/Y_{ij}) \cdots$	$Y_{mk} - Y_{ik}(Y_{mj}/Y_{ij})$
	$\cdots \quad Y_{in}/Y_{ij} \quad \cdots$	$Y_{tn} - Y_{in}(Y_{tj}/Y_{ij}) \cdots$	$\overline{M} + Y_{in}(Y_{mj}/Y_{ij})$

The pivoted tableaux are clearly dual. Notice especially that $M + \overline{M}$ stays constant in the pivoting, that is, the sum of the two objective values is the same after pivoting. Let the transforms of M and $\overline{M}$ under pivoting be $T(M)$ and $T(\overline{M})$ respectively. From the definition of dual tableaux, $M = -\overline{M}$ or $M + \overline{M} = 0$. Then

$$\begin{aligned} T(M) + T(\overline{M}) &= M - Y_{mj}(Y_{in}/Y_{ij}) + \overline{M} + Y_{in}(Y_{mj}/Y_{ij}) \\ &= M + \overline{M} \\ &= 0. \end{aligned}$$

Thus $T(M) = -T(\overline{M})$. When we have arrived at the feasible optimum for the dual tableau, $T(\overline{M})$ is a maximum and therefore $-T(\overline{M})$ is a minimum. Remember that in the dual program we replace the objective function to be minimized with the corresponding function $\overline{M}$ to be maximized, in order to use the same pivoting algorithm in both cases.

The next job is to show that when one tableau arrives at its feasible optimum, the dual tableau must also be at its feasible optimum. Suppose the pivoted primal tableau above is optimum and its objective maximum is $T(M)$. Then the first $n-1$ entries in its last row are positive, and since each pivot preserves feasibility the first $m-1$ entries in its right-hand column are likewise positive. Now the corresponding dual tableau must be feasible because its right-hand column contains respectively the $n-1$ positive numbers of the last row in the primal; and the dual tableau remains optimal since its last row contains respectively the $m-1$ positive numbers of the right-hand column in the primal. Thus the pivoting of the dual tableau is complete and $-T(\overline{M})$ is its objective minimum. But we have shown $T(M) = -T(\overline{M}) = m$, so that the maximum of the primal is the minimum of the dual. Finally this interchange of bottom row with right-hand column shows that both sets of numbers may be found from either one of the two optimized tableaux. ■

6.2 The Minimax Theorem

Theorem 6–2. (The Minimax Theorem) *The maximum value of the objective function in the solution to the primal program is the same as the minimum value of the objective function in the solution to the dual program.*

In the context of game theory the Minimax Theorem says that the maximum of A's minimum expectation is the same as the minimum of B's maximum expectation. In other words A's game value is the same as B's game value. Using only elementary techniques we have proved the von Neumann Minimax Theorem not only for zero-sum matrix games but also for general linear programs. We have in fact a choice of dual procedures for solving a linear program. We may solve the primal program by pivoting according to the Simplex Algorithm, or we may solve the dual program by pivoting according to the Dual Simplex Algorithm. It is wise to pick the method that leads to the simpler of the two solutions. While the two methods are equivalent to each other, one may be easier to set up than the other.

The symmetric dualization developed so far is appropriate for problems involving constraints that are all of type I or are all of type II. The situation for equality constraints and a mixture of types is more complicated. An equality is equivalent to a pair of inequalities, one of each type. Therefore, an equality constraint may be replaced with two inequality constraints, one of type I and the other of type II. This necessarily leads to a system of mixed constraints. Mixed systems will be considered in the next chapter.

Problems Chapter 6

6–1. For the following matrix game

$$\begin{bmatrix} 1 & 4 & 5 \\ 7 & 2 & 6 \\ 8 & 9 & 3 \end{bmatrix}$$

(a) Set up the dual linear programs for A's problem and B's problem.
(b) Pivot the condensed tableau for B and pivot by the Dual Simplex Method the condensed tableau for A verifying at each pivot the Fundamental Duality Theorem.
(c) Verify von Neumann's Minimax Theorem.
(d) State the optimal strategies and the value of the game.

6–2. Follow the same procedure in Problem 6–1 for the following matrix game.

$$\begin{bmatrix} 1 & -2 & 3 \\ -4 & 5 & 6 \end{bmatrix}$$

6–3. Solve the following linear program by first dualizing the problem and then pivoting. Carefully read the final tableau to obtain the complete solution vector. That is, state the values of the four slack variables as well as the three main variables of the original problem. As a check on your result convert the type II inequalities into type I, and then pivot by the Dual Simplex Algorithm.

$$\begin{aligned} x_i \geq 0,\ i = 1, 2, 3 \\ x_1 + 2x_2 \qquad\quad \geq 2 \\ 3x_1 + \ x_2 + \ x_3 \geq 4 \\ 4x_3 \geq 1 \\ x_1 \qquad\quad + 3x_3 \geq 1 \end{aligned}$$

where $4x_1 + 3x_2 + 3x_3 = m$ is a minimum.

6–4. For the following linear program carry out the same directions found in Problem 6–3.

$$\begin{aligned} x_i \geq 0,\ i = 1, 2, 3 \\ x_1 + 2x_2 \qquad\quad \geq 2 \\ 3x_1 + \ x_2 + \ x_3 \geq 4 \\ 4x_3 \geq 1 \\ 2x_1 \qquad\quad + \ x_3 \geq 1 \end{aligned}$$

where $4x_1 + 3x_2 + 3x_3 = m$ is a minimum.

6–5. Solve the following linear program and state the complete solution vector including the five slack variables.

$$
\begin{aligned}
x_1 \geq 0,\ x_2 &\geq 0 \\
3x_1 + 2x_2 &\geq 4 \\
2x_1 + x_2 &\geq 3 \\
-x_1 + 3x_2 &\geq 5 \\
-2x_1 + x_2 &\geq 2 \\
4x_1 + 2x_2 &\geq 5
\end{aligned}
$$

where $x_1 + x_2 = m$, a minimum.

6–6. Solve the following linear program and state the complete solution vector including all slack variables.

$$
\begin{aligned}
x_i \geq 0,\ i = 1, 2, 3 & \\
x_1 + 2x_2 &\geq 4 \\
3x_1 + x_2 + x_3 &\geq 5 \\
x_2 + 4x_3 &\geq 2 \\
x_1 + x_3 &\geq 3
\end{aligned}
$$

where $4x_1 + 3x_2 + 3x_3 = m$, a minimum.

7

PRIMAL-DUAL METHODS

7.1 Methods for Mixed Systems

The concept of duality not only has great theoretical significance but it also has considerable value in the solving of problems. The dual of a particular program may be much easier to solve than the original program and should be taken into consideration. By a combination of the Simplex Algorithm and the Dual Simplex Algorithm, artificial variables may be avoided completely. Avoiding artificial variables reduces the size of the tableau and usually makes a marked reduction in the number of iterations necessary to optimize the problem.

Let us consider the following example in two variables which may be graphed, and then the various solutions may be checked against the graph.

Example 1. Find $x_1 \geq 0$, $x_2 \geq 0$, satisfying constraints

$$\langle 1 \rangle \qquad x_1 + 7x_2 \leq 63$$

$$\langle 2 \rangle \qquad 3x_1 + x_2 \geq 9$$

$$\langle 3 \rangle \qquad 3x_1 + 2x_2 \geq 15$$

$$\langle 4 \rangle \qquad 5x_1 + 6x_2 \geq 30$$

$$\langle 5 \rangle \qquad x_1 + 4x_2 \geq 8$$

$$\langle 6 \rangle \qquad 8x_1 + 3x_2 \leq 80,$$

such that $2x_1 + x_2$ is a minimum.

The graph appears in Figure 7–1 in which the minimum is seen to occur at vertex (1, 6). The boundaries are labeled with the numbers of their corresponding constraints.

The given system of contraints contains four type II inequalities. Under the Simplex Method these will require four artificial variables. It is left as an exercise to solve the problem with artificial variables. Instead, let us label the six slack variables $x_3, \ldots, x_8$, multiply constraints $\langle 1 \rangle$ and $\langle 6 \rangle$ by -1, and then dualize. The symmetric system is

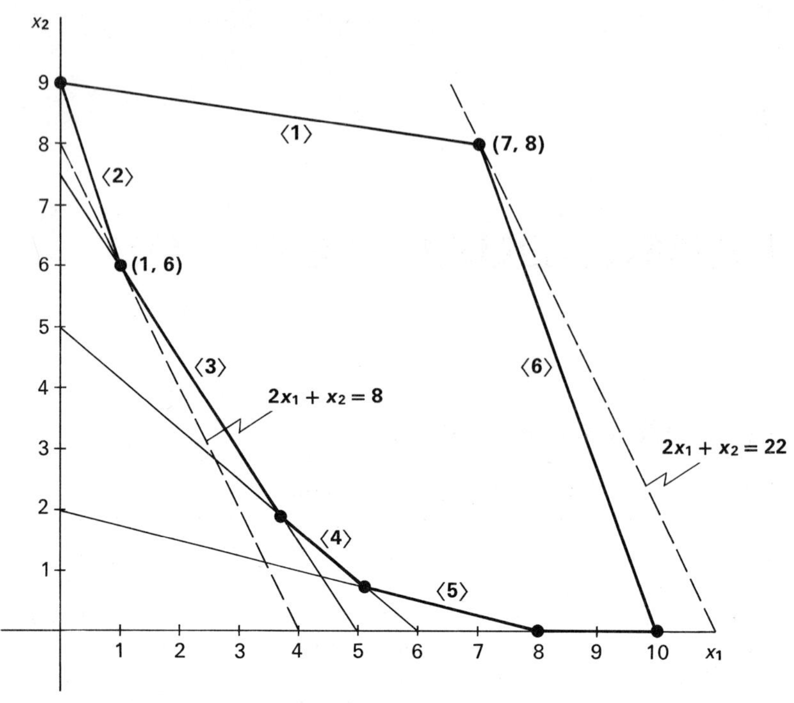

FIGURE 7–1

dualized by Definition 10 in Chapter 5 to give the following primal program.

Find $x_i \geq 0$, $i = 3, \ldots, 8$ satisfying

$$\langle 7 \rangle \qquad -1x_3 + 3x_4 + 3x_5 + 5x_6 + 1x_7 - 8x_8 \leq 2$$

$$\langle 8 \rangle \qquad -7x_3 + 1x_4 + 2x_5 + 6x_6 + 4x_7 - 3x_8 \leq 1$$

$$-63x_3 + 9x_4 + 15x_5 + 30x_6 + 8x_7 - 80x_8 = \overline{M},$$

where $\overline{M}$ is a maximum.

The von Neumann Theorem tells us that $\overline{M} = m$, the minimum of the dual program. Note that we are considering the problem as originally stated to be the dual. We therefore solve the primal program for its maximum by using the initial tableau of constraints ⟨7⟩ and ⟨8⟩ where x_1 and x_2 are the slack variables.

	3	4	5	6	7	8	
1	−1	3	3	5	1	−8	2
2	−7	1	2	6	4	−3	1
	63	−9	−15	−30	−8	80	0

This tableau may be run on our automatic routine from Chapter 4 which produces the final tableau listed next after three pivots.

	3	1	6	2	7	8	
4	6.33	.67	−2.67	−1	−3.33	−2.33	.33
5	−6.67	−.33	4.33	1	3.67	−.33	.33
	20	1	11	6	17	54	8

It is very important to read the final tableau correctly. Because we want the solution to the dual program, the basic variables are across the top of the final tableau with their corresponding values in the bottom row. The nonbasic variables are at the left. Thus the complete solution vector, including the slack variables, is

$$\begin{aligned} x_1 &= 1 \\ x_2 &= 6 \\ x_3 &= 20 \\ x_4 &= 0 \\ x_5 &= 0 \\ x_6 &= 11 \\ x_7 &= 17 \\ x_8 &= 54. \end{aligned}$$

The minimum value 8, in the lower right-hand corner, checks the value found from the objective function,

$$m = 2x_1 + x_2 = 2 + 6 = 8.$$

Note that x_4 and x_5 are zero since they are nonbasic. Only 3 pivots were required compared to 6 pivots to optimize the tableau with artificial variables. Also the tableau size has been reduced from 8×7 to 3×7.

An alternative to dualizing is to use the Dual Simplex Algorithm. Going back to the original constraints of Example 1, we multiply the four type II constraints by minus one. Then setting $m = -M$ will put the problem into the following form.

Maximize M subject to

$$\begin{aligned} x_1 \geq 0,\ x_2 &\geq 0 \\ x_1 + 7\,x_2 &\leq 63 \\ -3x_1 - \,x_2 &\leq -9 \\ -3x_1 - 2\,x_2 &\leq -15 \\ -5x_1 - 6\,x_2 &\leq -30 \\ -x_1 - 4\,x_2 &\leq -8 \\ 8x_1 + 3\,x_2 &\leq 80 \\ 2x_1 + \,x_2 + M &= 0. \end{aligned}$$

The original example has now been changed to a program with type I constraints for which the initial tableau is

	1	2	
3	1	7	63
4	−3	−1	−9
5	−3	−2	−15
6	−5	−6	−30
7	−1	−4	−8
8	8	3	80
	2	1	0

This tableau is precisely the dual of the previous initial tableau. It is optimal but not feasible. If we pivot by the Dual Simplex Algorithm the three pivots will be dual to the same three pivots of the dual tableau. For example, the θ ratios of the fourth row are $\theta_1 = -2/5$ and $\theta_2 = -1/6$. Since the numerically smallest ratio is θ_2, the first pivot will occur at (4, 2), fourth row, second column. The initial pivot in our previous solution was at (2, 4). The result after three pivots by the Dual Simplex Algorithm is the final tableau given below.

	4	5	
3	−6.33	6.67	20
1	−.67	.33	1
6	2.67	−4.33	11
2	1	−1	6
7	3.33	−3.67	17
8	2.33	.33	54
	.33	.33	−8

The two final tableaux are of course dual to one another. In this case $M = -8$, so the minimum $m = -M = 8$. ●

The two methods are equivalent but the first had the advantage of running on our automatic program of Chapter 4. The second must either be programmed by the Dual Simplex Algorithm or have the pivot position told to the machine after each tableau. The necessary subroutines for pivoting by the Dual Simplex Algorithm are found in Appendix A.

Let us now find the maximum of the same objective function over the given feasible region of example 1.

Example 2. From Figure 7–1 the maximum is seen to occur at vertex (7, 8) and thus $M = 2x_1 + x_2 = 22$. To find it by pivoting, we set up the following initial tableau which differs from the last initial tableau only in the bottom row.

	1	2	
3	1	7	63
4	−3	−1	−9
5	−3	−2	−15
6	−5	−6	−30
7	−1	−4	−8
8	8	3	80
	−2	−1	0

The present tableau is neither optimal nor feasible. The difficulty can be handled by a combination of our two methods. All possible pivots should be considered by both the Simplex and Dual Simplex Algorithms. In this case the Dual Simplex Algorithm produces no pivot since the last row is all negative. There are no possible negative θ ratios for determining a pivotal column. For the first column the Simplex Algorithm gives us the row ratios $\theta_1 = 63/1$ and $\theta_6 = 80/8$. The smaller ratio, θ_6, determines the pivot position to be (6, 1). After pivoting we have the following tableau.

	8	2	
3	−.125	6.625	53
4	.375	.125	21
5	.375	− .875	15
6	.625	−4.125	20
7	.125	−3.625	2
1	.125	.375	10
	.25	−.25	20

The second tableau is already feasible but not optimal. The next pivot determined by row ratios using the second column is at (1, 2). This leads to the final tableau below.

	8	3	
2	−.019	.151	8
4	.377	−.019	20
5	.358	.132	22
6	.547	.623	53
7	.057	.547	31
1	.132	−.057	7
	.245	.038	22

The maximum is 22 and the complete solution vector is read off as:

$$
\begin{aligned}
x_1 &= 7 \\
x_2 &= 8 \\
x_3 &= 0 \\
x_4 &= 20 \\
x_5 &= 22 \\
x_6 &= 53 \\
x_7 &= 31 \\
x_8 &= 0. \quad \bullet
\end{aligned}
$$

The same result is found using artificial variables but requires 7 pivots compared to the two pivots used here. Also the tableau size has been cut down from 8 × 7 to 7 × 3. So far we have had no conflicts in choosing a pivot. The possibility remains that a tableau is neither feasible nor optimal and the two methods each produce different pivots. A choice may be made by answering the question, "Which pivot makes the greater change in the objective value?" A pivot chosen by the Simplex Algorithm increases the objective value, advancing toward optimality. A pivot chosen by the Dual Simplex Algorithm decreases the objective value which both improves the objective of the dual and advances the primal toward feasibility.

To distinguish between the two types of θ ratios, let us use θ_i, $i = 1$ to $m - 1$, for the row ratios and $\bar{\theta}_j$, $j = 1$ to $n - 1$, for the column ratios. From Chapter 3 the formula for the objective value after a pivot at Y_{pq} is

$$
\begin{aligned}
\bar{Y}_{mn} &= Y_{mn} - Y_{mq}(Y_{pn}/Y_{pq}) \\
&= Y_{mn} - Y_{mq}\theta_p .
\end{aligned}
$$

The increase in the objective function is the absolute value of $Y_{mq}\theta_p$. Let

$$\langle 9 \rangle \qquad I_p = |Y_{mq}\theta_p|$$

be the increase in the objective function due to a pivot chosen by the use of row ratios. The corresponding formula for the decrease due to a pivot chosen by column ratios is

$$Y_{mq}Y_{pn}/Y_{pq} = \bar{\theta}_q Y_{pn}.$$

Let

$$\langle 10 \rangle \qquad D_q = |\bar{\theta}_q Y_{pn}|$$

be the decrease in the objective function due to a pivot chosen by the Dual Simplex Algorithm. When two pivots are possible the choice may be made by picking the pivot corresponding to the larger of I_p and D_q. Thus the next tableau will make the largest possible change in the program either toward optimality or toward feasibility. In most cases this choice should result in the fewest number of pivots necessary to reach a final tableau.

Let us solve the following illustrative example.

Example 3. Find $x_i \geq 0$, $i = 1, 2, 3, 4$, satisfying the constraints

$$x_1 - 2x_2 + 4x_3 - 3x_4 = 10$$
$$2x_1 + 3x_2 - 1x_3 + 5x_4 \leq 15$$
$$3x_1 - 1x_2 + 2x_3 + 3x_4 \geq 12,$$

where $x_1 + x_2 - 2x_3 - 4x_4 = M$ is a maximum.

The equality constraint is replaced with two inequalities, one of type I and one of type II. The type II inequalities are then multiplied by -1 to form a symmetrical maximizing problem. The equivalent system is

$$x_1 - 2x_2 + 4x_3 - 3x_4 \leq 10$$
$$-x_1 + 2x_2 - 4x_3 + 3x_4 \leq -10$$
$$2x_1 + 3x_2 - x_3 + 5x_4 \leq 15$$
$$-3x_1 + x_2 - 2x_3 - 3x_4 \leq -12$$
$$M - x_1 - x_2 + 2x_3 + 4x_4 = 0.$$

The slack variables are taken to be x_5, x_6, x_7 and x_8 respectively in the following initial tableau.

	1	2	3	4	
5	1	−2	4	−3	10
6	−1	2	−4	3	−10
7	2	3	−1	5	15
8	−3	1	(−2)	−3	−12
9	−1	−1	2	4	0

The initial tableau is neither feasible nor optimal. The possible row ratios for the first column are $\theta_1 = 10/1$ and $\theta_3 = 15/2$. Using the smaller ratio we compute

$$I_p = I_3 = |Y_{51}\theta_3| = 7.5.$$

The possible column ratios for the fourth row are $\bar{\theta}_3 = 2/-2$ and $\bar{\theta}_4 = 4/-3$. Choosing the smaller ratio in absolute value we find

$$D_q = D_3 = |\bar{\theta}_3 \, Y_{45}| = 12.$$

Since D_3 is larger than I_3 we pick the pivot according to the Dual Simplex Algorithm to be at (4, 3). The result of this pivot is to decrease M as shown in the next tableau.

	1	2	8	4	
5	−5	0	2	−9	−14
6	(5)	0	−2	9	14
7	3.5	2.5	−.5	6.5	21
3	1.5	−.5	−.5	1.5	6
9	−4	0	1	1	−12

The second tableau is still both infeasible and nonoptimal. We have reduced the infeasible variables to one even though $M = -12$ is further from a maximum. Once again computing the row ratios for the first column and the column ratios for the first row we find

$$I_2 = |Y_{51}\theta_2| = 4(14/5) = 11.2$$
$$D_4 = |\bar{\theta}_4 \, Y_{15}| = (1/9)14 = 1.56.$$

Since I_2 is the larger we pivot at (2, 1) to obtain the next tableau.

	6	2	8	4	
5	1	0	0	0	0
1	.2	0	−.4	1.8	2.8
7	−.7	2.5	(.9)	.2	11.2
3	−.3	−.5	.1	−1.2	1.8
9	.8	0	−.6	8.2	−.8

The third tableau is feasible but not yet optimal. The objective value has been increased to $M = -.8$. A final pivot is determined by the row ratios of the third column to be at (3, 3). Such a pivot must increase the objective value as shown in the final tableau.

	6	2	7	4	
5	1	0	0	0	0
1	−.111	1.111	.444	1.889	7.778
8	−.778	2.778	1.111	.222	12.444
3	−.222	−.778	−.111	−1.222	.556
9	.333	1.677	.667	8.333	6.667

The solution vector for the four main variables is

$$x_1 = 7.778$$
$$x_2 = 0$$
$$x_3 = .556$$
$$x_4 = 0.$$

For this solution the maximum of the objective function is $M = 6.667$ as may be checked. Slack variables x_5 and x_6 are necessarily both zero since they arose from an equality constraint. The final tableau appears to be degenerate with a basic variable equal to zero. However, no further pivoting is possible and so the solution is unique. Slack variables associated with an equality constraint must be zero in the final tableau and do not represent a true degeneracy in the optimal basic solution. ●

The solution to this example may be achieved in 4 pivots by use of artificial variables.

7.2 The Primal-Dual Algorithm

In each previous example a worthwhile savings was realized by the combined methods both in the number of iterations necessary and in tableau size. Let us summarize our technique in algorithm form.

The Primal-Dual Algorithm for Mixed Systems

1. Replace equality constraints with two inequalities, one of type I and the other of type II.
2. Convert all type II inequalities to type I by multiplication of -1.
3. If necessary, convert the objective function to a maximization objective.
4. Set up the initial tableau with a slack variable in the basis for each constraint. Eventually all variables are to satisfy the nonnegativity requirement.
5. For the most negative number in the last column determine a possible pivot in that row by column ratios according to the Dual Simplex Algorithm. The ratio must be negative. If no pivot is available, move to the row of the next most negative number in the last column and try again. Continue through the negatives of the last column until either a possible pivot is found or none is available.
6. For the most negative number in the bottom row, determine a possible pivot in that column by row ratios as in the Simplex Algorithm. The ratio must be positive. If no pivot is available, continue to try the remaining columns with a negative in the bottom row until either a pivot is found or none is available.
7. If steps 5 and 6 each produce a pivot, compute I_p and D_q by formulas $\langle 9 \rangle$ and $\langle 10 \rangle$ respectively in Section 7.1. Choose the pivot corresponding to the larger, I_p or D_q.
8. After pivoting repeat steps 5 through 7 until the tableau is optimal and feasible or until no new pivot can be found.

A correct interpretation of the final tableau is of prime importance. The final tableau may be optimal and feasible. In this case, a solution has been found to both the primal and dual programs. The solution may be degenerate but true degeneracy should be distinguished from that arising when equality constraints are present. When an equality constraint is satisfied, the two associated slack variables are reduced to zero. One of these slacks is driven out of the basis but the other remains in the solution at value zero. This is not a true degeneracy as was noted in our last example. If the final tableau is optimal but contains a negative in the last column due to the fact that no pivot can be found in that row, then the primal program has no feasible solutions and its dual is unbounded. If the final tableau is feasible but contains a negative in the objective row for which all other entries in that column are nonpositive, then the primal objective is unbounded and the dual program has no feasible solutions. The final tableau may also be both infeasible and nonoptimal. For an example see Problem 7–22. One of the above situations will necessarily occur so that the original problem either has an optimal solution, or it is unbounded, or it is inconsistent, or it is both unbounded and inconsistent.

In reading off the answers from the final tableau, care must be taken to distinguish between the basic and nonbasic variables. For the primal program the basic variables are denoted by the subscripts $L(I)$ to the left of our tableau, and their values are found in the right-hand column. For the dual program the basic variables are denoted by the subscripts $K(J)$ across the top of our tableau, and their corresponding values are found in the bottom row. In both cases the nonbasic variables are always zero.

A set of subroutines in the FORTRAN language for carrying out the Primal-Dual Algorithm is found in Appendix A.

7.3 Shadow Prices

One of the important interpretations of duality is discovered in the economic meaning of the dual solution. A manufacturer is interested in knowing the highest price he can afford to pay for additional material to increase his output. The question of whether or not it is profitable to buy raw material at a certain cost may be answered by what is called the Shadow Price of that commodity. The idea will be illustrated in the following example.

Example 4. A women's apparel manufacturer has a production line making two styles of girls' skirts. Style one is a miniskirt which requires 2 ounces of cotton thread, 3 ounces of dacron thread, and 3 ounces of linen thread. Style two is a micro-miniskirt which requires 2 ounces of cotton thread, 2 ounces of dacron thread, and 1 ounce of linen thread. The manufacturer realizes a net profit of \$1.95 on style one and a net profit of \$1.59 on style two. He has on hand an inventory of 15 pounds of cotton thread, 16 $\frac{1}{4}$ pounds of dacron thread, and 13 $\frac{3}{4}$ pounds of linen thread. His immediate problem is to determine a production schedule, given the current inventory, to make a maximum profit. Then he would like to know at what prices per oz. would it be profitable to buy more thread, assuming a small increase in production is desired.

Solution. We will first solve the immediate problem. Let x_1 be the number of miniskirts produced and x_2 the number of micro-miniskirts produced with the current inventory. Expressing the quantities of thread in ounces, we arrive at the following constraints:

$$2x_1 + 2x_2 \leq 240$$

$$3x_1 + 2x_2 \leq 260$$

$$3x_1 + 1x_2 \leq 220.$$

The objective is to maximize the profit,

$$M = 1.95\,x_1 + 1.59\,x_2\,.$$

Let x_3 be the slack ounces of cotton, x_4 be the slack ounces of dacron, and x_5 be the slack ounces of linen. The initial tableau is

	1	2	
3	2	2	240
4	3	2	260
5	3	1	220
6	−1.95	−1.59	0

After three pivots, we reach the final tableau of the primal program

	3	4	
5	1.5	−2	60
2	1.5	−1	100
1	−1	1	20
6	.435	.36	198

With the thread on hand, the manufacturer should produce $x_1 = 20$ miniskirts and $x_2 = 100$ micro-miniskirts at a total profit of \$198. By the Duality Theorem, this optimum value may be computed either from the objective function of the primal program or the objective function of the dual program. These two objective functions are given below in equations $\langle \mathbf{1} \rangle$ and $\langle \mathbf{2} \rangle$ where the x_i's are read off of the final tableau.

$$\langle \mathbf{1} \rangle \qquad M = 1.95\, x_1 + 1.59\, x_2$$
$$= 1.95\,(20) + 1.59\,(100) = \$198.$$

$$\langle \mathbf{2} \rangle \qquad m = 240\, x_3 + 260\, x_4 + 220\, x_5$$
$$= 240\,(.435) + 260\,(.36) + 220\,(0) = \$198.$$

In the dual problem, each of the variables must be given an appropriate economic meaning that is quite different from their definition in the primal problem. Variables x_3, x_4, and x_5 may be thought of as the cost per ounce of cotton, dacron, and linen respectively. The dual objective is then to find x_3, x_4, and x_5, so that these unit costs will minimize equation ⟨**2**⟩. Of course, the production schedule that maximizes profit on skirts is the same as the production schedule that minimizes the total cost of raw material. These unit costs, $x_3 = .435$, $x_4 = .36$, and $x_5 = 0$, also represent an upper bound on the amount the manufacturer should pay for additional raw material. For example, if he buys an additional ounce of cotton, equation ⟨**2**⟩ shows that his profit is increased by 43 $\frac{1}{2}$ cents. Likewise, an additional ounce of dacron would increase profit by 36 cents. ●

Definition. The **shadow price** of a commodity is the unit price of that commodity that is equal to the increase in profit to be realized by one additional unit of that raw material.

Naturally the shadow price is the highest price that the manufacturer would be willing to pay for more raw material. He wishes to buy raw material at less than its shadow price in order that the corresponding increase in output will also increase net profits.

The shadow prices of cotton and dacron from Example 4 are $43\frac{1}{2}$ cents per ounce and 36 cents per ounce respectively. The shadow price of linen is zero since it does not increase profit to buy more linen until the current supply is exhausted. In the basic solution to the primal program, x_5 shows us that there are 60 ounces of slack or unused linen. In general, if a slack variable is nonzero in the solution, then its shadow price will be zero. Another popular term for shadow price is **marginal price**.

Equation ⟨**2**⟩ may be used to find the optimum of a new problem without resolving provided the basic variables don't leave the basis. We have already noted in Example 4 that if an additional ounce of dacron is available, the new profit would be \$198.36. As a check, let us solve Example 4 with 261 ounces of dacron. The final tableau reached in 3 pivots is

	3	4	
5	1.5	−2	58
2	1.5	−1	99
1	−1	1	21
6	.435	.36	198.36

The only change is in the last column where the solution requires the production of one more miniskirt and one less micro-miniskirt for the new profit of \$198.36.

The question of how much a resource can be changed while the shadow prices remain the same is a problem in Sensitivity Analysis. Geometrically, this problem may be viewed as, how much can the coefficients of the dual objective function be changed without slipping off of the optimal vertex in the dual solution? If the dual program can be graphed, the answer is seen in considering the slopes involved. In Example 4 the dual program may be stated as follows. Find x_3, the cost per ounce of cotton thread, x_4, the cost per ounce of dacron thread, and x_5, the cost per ounce of linen thread, satisfying

$$\langle 3 \rangle \qquad \begin{aligned} 2x_3 + 3x_4 + 3x_5 &\geq 1.95 \\ 2x_3 + 2x_4 + x_5 &\geq 1.59, \end{aligned}$$

such that $240\, x_3 + 260\, x_4 + 220\, x_5 = m$ is the minimum total cost of raw material.

In the dual objective function the ratio of cotton to dacron is $\frac{12}{13}$. The corresponding ratios in the two constraints $\langle 3 \rangle$ are $\frac{2}{3}$ and $\frac{2}{2} = 1$. This suggests that in the objective function, cotton could be increased by 20 ounces to 260 ounces so that its ratio to dacron is one. Up to an increase of 20 ounces of cotton, the objective plane will intersect the same optimal vertex. At this point the solution becomes degenerate, but the optimum is still given by

$$\begin{aligned} m &= 260\,(.435) + 260\,(.36) + 0 \\ &= 113.10 + 93.60 = \$206.70. \end{aligned}$$

We can verify this minimum by solving Example 4 with the resource of cotton at 260 ounces. The result is given in the next tableau reached after three pivots.

	3	4	
5	1.5	−2	90
2	1.5	−1	130
1	−1	1	0
6	.435	.36	206.70

The optimum now requires the production of 130 micro-miniskirts and no miniskirts.

Suppose only the dacron is increased. The increase is limited to 30 ounces because at that point the slack linen is used up and the solution again becomes degenerate. Let us check the answers for an increase of 29 to 289 ounces of dacron. The solution using our shadow prices is immediately

$$\begin{aligned} m &= 240(.435) + 289(.36) + 0 \\ &= 104.40 + 104.04 = \$208.44. \end{aligned}$$

The verification of this minimum by solving Example 4 with the resource of dacron at 289 ounces is seen in the following final tableau.

	3	4	
5	1.5	−2	2
2	1.5	−1	71
1	−1	1	49
6	.435	.36	208.44

From this final tableau the optimum production schedule is 49 miniskirts and 71 micro-miniskirts. In each of the variations of Example 4, the maximum profit was computed from the shadow prices without resorting to the solution of the new program. A variety of initial conditions was possible under the same shadow prices. A further look at Sensitivity Analysis will be made in Chapter 9 with the use of a parameter in the resource column.

Problems Chapter 7

7–1. Solve the following linear program in several ways considering both primal and dual methods. Check your result graphically. State the optimum value and the entire solution vector including slack variables.

$$\begin{aligned} x_1 \geq 0,\ x_2 &\geq 0 \\ 2x_1 + \ x_2 &\geq 6 \\ 2x_1 + 3x_2 &\geq 10 \\ -3x_1 + 5x_2 &\geq -24 \\ 5x_1 + 2x_2 &\leq 71 \\ 2x_1 + 7x_2 &\leq 78 \\ 3x_1 - 4x_2 &\geq -28, \end{aligned}$$

where $5x_1 + 3x_2 = M$ is a maximum.

7–2. Carry out Problem 7–1 where the objective is to minimize $5x_1 + 3x_2$.

7–3. Solve the following problem without using artificial variables and note the values of all slack variables.

$$\begin{aligned} x_1 \geq 0,\ x_2 &\geq 0 \\ 2x_1 + \ x_2 &\leq 35 \\ x_2 &\leq 13 \\ 4x_1 + \ x_2 &\geq 12 \\ 2x_1 + \ x_2 &\geq 10 \\ x_1 + \ x_2 &\geq 7 \\ x_1 + 4x_2 &\geq 10 \\ x_1 - \ x_2 &\leq 10, \end{aligned}$$

where $3x_1 + x_2 = M$ is a maximum.

7–4. Find the minimum value of the objective function in Problem 7–3 along with its solution vector.

7–5. Find both the maximum and the minimum value of $2x_1 - 3x_2$ under the same constraints given in Problem 7–3 without using artificial variables.

7–6. Solve Problem 7–5 with artificial variables and compare the number of pivots needed with the number used in 7–5.

7–7. Find the solution vector and the minimum value of m for

$$\begin{aligned} x_i \geq 0,\ i &= 1, 2, 3 \\ x_1 + 2x_2 - \ x_3 &\leq 9 \\ 2x_1 - \ x_2 + 2x_3 &= 4 \\ -x_1 + 2x_2 + 2x_3 &\geq 5 \\ 2x_1 + 4x_2 + \ x_3 &= m. \end{aligned}$$

7–8. Find nonnegative numbers x_i, $i = 1, \ldots, 5$, and the maximum, M, such that

$$
\begin{aligned}
2x_1 + 3x_2 - 4x_3 + 3x_4 - 2x_5 &= 2 \\
3x_1 - 2x_2 + 5x_3 - 4x_4 + 3x_5 &= 7 \\
x_1 + 2x_2 - 3x_3 - 3x_4 + 2x_5 &= M.
\end{aligned}
$$

7–9. Does the objective function in Problem 7–8 possess a minimum under the same constraints?

7–10. Find nonnegative numbers x_i, $i = 1, 2, 3$, and the minimum m, such that

$$
\begin{aligned}
3x_1 + 2x_2 + 4x_3 &\leq 26 \\
x_1 + 2x_2 + x_3 &\geq 13 \\
3x_1 + x_2 + 2x_3 &= 16 \\
x_1 + 2x_2 + x_3 &= m.
\end{aligned}
$$

7–11. Find nonnegative numbers x_i, $i = 1, 2, 3$, and the minimum, m, such that

$$
\begin{aligned}
3x_1 + 2x_2 + 4x_3 &\leq 26 \\
2x_1 + x_2 + 2x_3 &\geq 13 \\
2x_1 + 5x_2 + 3x_3 &\geq 17 \\
3x_1 + x_2 + 2x_3 &= 16 \\
x_1 + 2x_2 + x_3 &= m.
\end{aligned}
$$

What are the values of the three slack variables?

7–12. Solve the following problem.

$$
\begin{aligned}
x_i \geq 0,\ i &= 1, 2, 3, 4 \\
5x_1 - 3x_2 + 2x_3 + 4x_4 &\leq 21 \\
-6x_1 + 2x_2 - 3x_3 + 3x_4 &= 1 \\
4x_1 + 5x_2 + 6x_3 - 5x_4 &\geq 12 \\
x_1 + x_2 + x_3 + x_4 &\leq 10,
\end{aligned}
$$

where $x_1 + 2x_2 - 3x_3 + x_4 = M$, a maximum.

7–13. Solve Problem 7–12 where the objective function is to be minimized.

7–14. Solve the following problem where x_3 is unrestricted in sign, that is, x_3 may be positive, negative, or zero.

$$
\begin{aligned}
x_1 \geq 0,\ x_2 &\geq 0 \\
3x_1 + 5x_2 + 4x_3 &\leq 41 \\
2x_1 - 6x_2 - 9x_3 &\leq 11 \\
-x_1 + 3x_2 + 2x_3 &\leq 7,
\end{aligned}
$$

where $-9x_1 + 30x_2 - 40x_3 = M$, a maximum.

Hint: Replace x_3 with two variables, say y_3 and y_4, such that $x_3 = y_3 - y_4$ where both y_3 and y_4 are nonnegative. Then x_3 may be negative if $y_3 < y_4$, positive if $y_3 > y_4$, or zero if $y_3 = y_4$. Notice that this adds a new column to the initial tableau that is precisely the negative of the column corresponding to the original coefficients of x_3.

7–15. Show that the values of the two variables suggested in Problem 7–14 for replacing an unrestricted variable, x_i, are uniquely determined by the value of x_i. Hint: You must show that the two y variables cannot both be in the basis at the same time. Thus at least one of the y variables is always zero.

7–16. Solve the following problem where x_1 is unrestricted in sign,

$$\begin{aligned} &x_2 \geq 0,\ x_3 \geq 0 \\ &5x_1 + \ 3x_2 + \ 2x_3 = 2 \\ &6x_1 + \ 7x_2 + \ 4x_3 = 12 \\ &10x_1 + 15x_2 + 20x_3 = M, \text{ a maximum.} \end{aligned}$$

7–17. Find x_i, $i = 1, 2, 3, 4$, and the minimum value of m where x_2 and x_4 are unrestricted in sign and

$$\begin{aligned} &x_1 \geq 0, \qquad x_3 \geq 0 \\ &5x_1 + 9x_2 + 2x_3 - 2x_4 \leq 19 \\ &8x_1 - 3x_2 \qquad\quad - 4x_4 \geq 9 \\ &7x_1 - 2x_2 + \ x_3 - \ x_4 \leq 3 \\ &2x_1 + 5x_2 - \ x_3 \qquad\quad \geq 3 \\ &\ x_1 + \ x_2 - \ x_3 + \ x_4 = m. \end{aligned}$$

7–18. The Sunnydale Dairy processes 50,000 gallons of raw milk daily. In order to meet the needs of its regular customers at least 30,000 gallons will be bottled as fresh milk, at least 6000 gallons will be processed into cheese and at least 3000 gallons will be used for butter. The manager of Sunnydale will allow at most 4000 additional gallons to be bottled for fresh milk sales to other customers. The processing equipment for cheese and butter can handle no more than a total of 15,000 gallons of raw milk daily. The remainder of the daily supply will be made into powdered milk. However, the manager requires that at least 5000 gallons go to the powdered milk department in order to keep the drying equipment busy. The gross profit from sales figured per gallon of raw milk used in the processing is 50¢ on fresh bottled milk, 10¢ on cheese, 12¢ on butter, and 11¢ on powdered milk. What production schedule subject to these conditions will yield a maximum gross profit to Sunnydale Dairy? What is the shadow price of raw milk that goes into fresh bottled milk? In this case the shadow price represents the increase in profit possible for each additional gallon that the manager allows to go into fresh milk assuming that it can be sold.

7–19. Ace Advertising Agency has three principal promotion schemes that are carried out by two writers, a designer, a promoter, and a typist. The following table gives the time required in hours for schemes 1, 2, and 3 by each employee and their hours available per month.

	schemes			hours available
	1	2	3	
writer A	6	10	9	180
writer B	2	0	8	90
designer	8	6	10	176
promoter	12	5	8	180
typist	4	6	10	160

The three schemes sell for $500, $400, and $1000 respectively. How many customers for each type of scheme are needed per month in order to earn a maximum income? What are the shadow prices per hour of time associated with each employee? Which of the employees would you be willing to hire for overtime to increase company income and what would be a reasonable sum to pay for their overtime?

7–20. Two products from the Stoker manufacturing plant each require the three operations of casting, milling, and machining. The castings for the two products may be bought or made locally. If purchased, casting A costs $4 and casting B costs $8.90. The remaining costs, selling prices, and times involved for products A and B are given in the following table.

	A	B
casting made	$3.50	$7.50
milling	$12.00	$6.00
machining	$2.40	$5.00
selling price	$19.90	$22.45
milling time	1 hr.	$\frac{1}{2}$ hr.
machining time	12 min.	25 min.

If made at the local plant, casting A uses 3 lbs. of material C and 1 lb of material D, while casting B uses 5 lbs. of material C and 2 lbs. of material D. There are 8000 lbs. of C and 4000 lbs. of D available for the current two month period. The casting equipment at Stoker's can cast at most 1500 units of A or B during this two months. The milling machinery is available for a total of 1464 hours and the machining equipment for a total of 704 hours during the period. What should be the two month production schedule, and how many of the castings should be bought instead of made? What are the shadow prices of C and D, casting units, hours of milling time, and minutes of machining time? What resources need to be increased for an immediate increase in production?

7–21. The Nutrition Problem was published in 1945 by George J. Stigler before the advent of linear programming. At that time the solution was by trial and error because no direct method was available. Solve the following version by linear programming. A consumer desires at least the following quantities of nutriment per day.

proteins	100 gm.
calories	4000
calcium	800 mg.
iron	12 mg.

He decides that his diet will be made up from brown bread, butter, baked beans, cheddar cheese, and spinach. The consumer prices these articles per 100 grams at the local market and then looks up their ingredients in a table of the chemical composition of foods. The results of his research per 100 grams of food are shown in the following table.

	gm. protein	calories	mg. calcium	mg. iron	price
brown bread	8.3	246	17.2	2.01	$.07
butter	.4	793	14.8	.16	.21
baked beans	6.	93	61.6	2.05	.075
cheddar cheese	24.9	423	810.	.57	.19
spinach	5.1	26	595.	4.	.07

What should the consumer purchase in order to satisfy his nutritional requirements and pay the minimum cost per day? What is his minimum cost? Find the shadow prices of protein, calories, calcium, and iron.

7–22. Dantzig has given the following example in which both the problem and its dual are infeasible. Show that the final tableaux of this problem and its dual are both infeasible and unbounded.

$$x_1 \geq 0,\ x_2 \geq 0$$
$$x_1 - x_2 \geq 5$$
$$x_1 - x_2 \geq -5$$
$$-x_1 - x_2 = m,$$

where m is a minimum.

8

INTEGER PROGRAMMING

8.1 Introduction

Many problems require integral solutions because we cannot buy, sell or utilize a fractional unit of some product. Unfortunately, the best integer solution is not always the rounded off value of a general solution. We need a method of examining feasible points that have integral coordinates but are not necessarily vertices of the original feasible region. Such a method has been devised by R. E. Gomory. In order to follow this method we need a brief introduction into congruences of numbers.

8.2 Congruence

Let us first define congruence between rational numbers and then see how this idea allows us to represent whole classes of numbers with a single number.

Definition 1. For a pair of rational numbers, a and b, ***a* is congruent to *b*** if the difference $a - b$ is a multiple of some integer called the **modulus**.

We will use three bars to stand for *is congruent to*, a notation introduced by the great Carl F. Gauss. Then a is congruent to b modulo n is written $a \equiv b \pmod{n}$ which means $a - b = kn$ where k is an integer. Some examples are as follows.

$$10 \equiv 1 \pmod{3} \qquad -\tfrac{1}{2} \equiv \tfrac{5}{2} \pmod{3}$$

$$\tfrac{17}{5} \equiv -\tfrac{3}{5} \pmod{4} \qquad -7 \equiv 1 \pmod{4}$$

The congruence relationship with respect to any modulus is reflexive, symmetric, and transitive. These three properties, $a \equiv a$(reflexive), $a \equiv b$ implies $b \equiv a$(symmetric), and the two statements $a \equiv b$, $b \equiv c$ imply $a \equiv c$(transitive) are easily verified. In the case of transitivity note that $a - c = (a - b) + (b - c)$ so that if the latter two are multiples of some modulus so is the former. Any relation that satisfies these three laws is called an **equivalence relation** and numbers equivalent to one another are said to belong to the same **equivalence class**. All numbers congruent to one another belong to the same equivalence class.

Integer programming is based on the modulus 1. Two rational numbers are congruent modulo 1 if and only if their difference is an integer. Given a rational number a we wish to find another rational number b that is congruent to a modulo 1. Of course there are many possibilities. In order to narrow the field down we will restrict the second number b to lie in the interval $0 \leq b < 1$. For convenience let us define a special function to indicate this relationship.

Definition 2. For any rational number a, **function** f is defined by $f(a) = b$, where $a - b$ is an integer and b lies in the interval $0 \leq b < 1$.

Some examples of this notation follow:

$$f(\tfrac{1}{7}) = f(\tfrac{43}{7}) = f(-\tfrac{13}{7}) = f(-\tfrac{6}{7}) = \tfrac{1}{7}$$
$$f(-\tfrac{3}{4}) = f(\tfrac{17}{4}) = f(-\tfrac{7}{4}) = f(\tfrac{1}{4}) = \tfrac{1}{4}$$
$$f(5) = f(-9) = f(n) = 0, \text{ for any integer } n.$$

Theorem 8–1. For any rational number a, the number b, such that $f(a) = b$, is unique.

Proof. Suppose there are two such numbers b for some rational number a; that is, $f(a) = b_1$, $0 \leq b_1 < 1$ and $f(a) = b_2$, $0 \leq b_2 < 1$. Then by definition

$$a - b_1 = k_1$$
$$a - b_2 = k_2,$$

where k_1 and k_2 are integers. Subtracting these two equations gives $b_2 - b_1 = k_1 - k_2 = k$, where k is an integer. Since both b_1 and b_2 are nonnegative and less than one, their difference must be strictly between -1 and 1. The only integer in this range is $k = 0$. Therefore, $b_2 = b_1$ and $f(a)$ is unique. ■

Since $f(a)$ determines a unique rational number, we give it a special name.

Definition 3. The number b, determined by $f(a) = b$, is called the **congruent equivalent** to a, that is,

$$a \equiv b \pmod{1}$$

where b is a nonnegative fraction less than 1.

Some further properties of the new notation are given in the theorems below. Congruences will be understood to be (mod 1).

Theorem 8–2. $f(a + b) \equiv f(a) + f(b)$

Proof. Let

$$f(a+b) = c, \ 0 \leq c < 1,$$
$$f(a) = d, \ 0 \leq d < 1,$$

and

$$f(b) = e, \ 0 \leq e < 1.$$

We wish to show that $c \equiv d + e$. By the definition of f,

$$(a+b) - c = k_1$$
$$a - d = k_2$$
$$b - e = k_3,$$

where k_1, k_2, k_3 are integers. Subtracting the bottom two equations from the first equation gives

$$d + e - c = k_1 - k_2 - k_3 = k, \text{ an integer.}$$

Thus $d + e \equiv c$. ■

Theorem 8–3. If k is an integer then $f(ka) \equiv kf(a)$.

Proof. Let $f(ka) = c$, $0 \leq c < 1$, and $f(a) = d$, $0 \leq d < 1$. We wish to show $c \equiv kd$. By the definition of f,

$$ka - c = k_1$$
$$a - d = k_2,$$

where k_1 and k_2 are integers. Multiplying the second equation by k and subtracting from the first gives

$$kd - c = k_1 - kk_2 = k_3, \text{ an integer.}$$

Thus $kd \equiv c$. ■

Note that if k is not an integer in Theorem 8–3 then it cannot in general be factored out. For example, $f(\frac{1}{2} \cdot \frac{4}{3}) \neq \frac{1}{2}f(\frac{4}{3})$ since $f(\frac{2}{3}) = \frac{2}{3}$ and $\frac{1}{2}f(\frac{4}{3}) = \frac{1}{2} \cdot \frac{1}{3} = \frac{1}{6}$.

Theorem 8–4. If $a \equiv b$ then $f(a) = f(b)$, and conversely, if $f(a) = f(b)$ then $a \equiv b$.

Proof. If $a \equiv b$ then $a - b = k_1$, an integer. Let $f(a) = c$, $0 \le c < 1$, and let $f(b) = d$, $0 \le d < 1$. We must show $c = d$. From the definition of f,

$$\begin{aligned} a - c &= k_2 \\ b - d &= k_3, \end{aligned}$$

where k_2 and k_3 are integers. Subtracting the second equation from the first gives

$$\begin{aligned} a - b + d - c &= k_2 - k_3 \\ k_1 + d - c &= k_2 - k_3 \\ d - c &= k_2 - k_3 - k_1 = k, \end{aligned}$$

where k is an integer. As in the proof of Theorem 8–1, the only integer in the range of values for $d - c$ is zero. Thus $k = 0$ and $d = c$. Conversely, if $f(a) = f(b)$ then $f(a) - f(b) = 0 \equiv 0$. Using Theorem 8–2, Theorem 8–3, and transitivity, we have

$$\begin{aligned} f(a - b) &= f(a + (-b)) \\ &\equiv f(a) + f(-b) \\ &\equiv f(a) - f(b) \equiv 0. \end{aligned}$$

To be congruent to zero, $f(a - b)$ must be an integer, but the only integral value of f is zero. Thus $f(a - b) = 0$ and $a - b$ must be an integer. This completes the proof that $a \equiv b$. ■

With the use of Theorems 8–1 and 8–4, we now have the rational numbers divided up into equivalence classes (mod 1) so that the congruent equivalent, $f(r)$, for any member r of an equivalence class, is the unique representative of the entire class.

8.3 Gomory's Algorithm

Returning to the problem of finding integral solutions, let us suppose we have an optimal feasible tableau with a nonintegral basic variable. Let x_r be the basic variable that we wish to make integral. Suppose x_r is in the ith row of our optimal tableau, and let Y_{ij}, $j = 1, 2, \ldots, n$, be the tableau entries across that row. The current basic solution involves $x_r = Y_{in}$ where Y_{in} is not integral. If the nonbasic variables are $x_s, x_t, \ldots, x_u$ then the ith row corresponds to the equations

$$x_r + Y_{i1}x_s + Y_{i2}x_t + \cdots + Y_{in-1}x_u = Y_{in}$$

$$\langle 1 \rangle \qquad x_r = Y_{in} - (Y_{i1}x_s + Y_{i2}x_t + \cdots + Y_{in-1}x_u).$$

In order to make x_r integral, we want the right-hand side of equation ⟨**1**⟩ to be congruent to zero, modulo one understood. Of course this condition is currently not satisfied because the nonbasic variables $x_s, x_t, \ldots, x_u$ are all zero. Let us think of these x_j's as unknowns to be brought into the solution by pivoting. Then the condition ⟨**2**⟩ below will act as an unsatisfied constraint on the problem that will force x_r to be integral when satisfied.

$$\langle \mathbf{2} \rangle \qquad Y_{in} - (Y_{i1}x_s + Y_{i2}x_t + \cdots + Y_{in-1}x_u) \equiv 0.$$

From the definition of congruence, ⟨**2**⟩ is the same as

$$Y_{in} \equiv Y_{i1}x_s + Y_{i2}x_t + \cdots + Y_{in-1}x_u.$$

By Theorem 8–4 we have

$$f(Y_{in}) = f(Y_{i1}x_s + \cdots + Y_{in-1}x_u).$$

Using Theorem 8–2 and transitivity

$$\langle \mathbf{3} \rangle \qquad f(Y_{in}) \equiv f(Y_{i1}x_s) + \cdots + f(Y_{in-1}x_u).$$

We eventually want all **basic variables** to be integral so we further constrain ⟨**3**⟩ by considering only integral values of x_j. In this case the x_j may be factored out according to Theorem 8–3 giving

$$\langle \mathbf{4} \rangle \qquad f(Y_{in}) \equiv f(Y_{i1})x_s + \cdots + f(Y_{in-1})x_u.$$

The right- and left-hand sides of ⟨**4**⟩ are in the same equivalence class and so differ by an integer. Let

$$f(Y_{in}) = \phi, \ 0 < \phi < 1.$$

Then the right-hand side of ⟨**4**⟩ is $\phi + k$, for k a nonnegative integer. Thus ⟨**4**⟩ may be interpreted as a linear constraint with type II inequality as shown in ⟨**5**⟩.

$$\langle \mathbf{5} \rangle \qquad f(Y_{i1})x_s + \cdots + f(Y_{in-1})x_u \geq f(Y_{in}).$$

Furthermore the slack variable for ⟨**5**⟩, call it x_v, has the integral value $x_v = k$. Subtracting x_v from the left side of ⟨**5**⟩ and multiplying by (-1) gives the new equation to be added to our optimal tableau.

$$\langle \mathbf{6} \rangle \qquad x_v - f(Y_{i1})x_s - \cdots - f(Y_{in-1})x_u = -f(Y_{in}).$$

After adding in equation ⟨**6**⟩ as a new row, the tableau will be infeasible but remains optimal. Now pivoting can continue by the Dual Simplex Method.

The result of pivoting is to bring new variables into the basis until the new constraint is satisfied. If these new variables come in at integral levels, then the steps in the argument above can be reversed so that congruence ⟨**4**⟩ implies ⟨**3**⟩ implies ⟨**2**⟩. Thus, x_r will be integral. After pivoting terminates, if x_r is not yet integral or if nonintegral basic variables remain, the process can be repeated by adding new constraints similar to ⟨**5**⟩. Eventually both the slack and main variables will be driven to integral values. For a proof of actual convergence of this method in a finite number of steps you are referred to Gomory's original papers.[1] The proof assumes that the objective function has a lower as well as an upper bound. If the problem is one of maximizing, then the general solution provides an upper bound for all integer solutions. The existence of a lower bound for integer solutions is guaranteed if the feasible region is bounded. All that is necessary if an integer solution exists is to know that the corresponding value of the objective function is greater than some negative constant. It suffices to say that the convergence is fairly rapid for small problems and usually only a few applications of the algorithm are needed. For large problems other techniques are available that converge more rapidly.[2]

Each new constraint that is added to the optimal tableau is called a **cutting plane**. A cutting plane removes a part of the feasible region not containing points with integral coordinates. The cuts create new vertices until an optimal integral vertex is found. Since no integral solutions are removed, an optimal solution of the reduced feasible region is an optimal integral solution of the original problem. In starting the algorithm a choice must be made among the rows of the optimal tableau for the one to be used in constructing a cutting plane. There is no sure fire way of making this choice. Experience has shown that it is usually quickest to pick the row with the largest fractional part in its right-hand column, that is, for which $f(Y_{in})$ is the greatest. We summarize the steps in algorithm form.

Gomory's Algorithm

1. Pick the row from the optimal feasible tableau for which $f(Y_{in})$ is the largest to form the constraint

$$f(Y_{i1})x_s + \cdots + f(Y_{in-1})x_u \geq f(Y_{in}).$$

2. Subtract a slack variable, x_v, and multiply both sides by (-1) to form the equation

$$x_v - f(Y_{i1})x_s - \cdots - f(Y_{in-1})x_u = -f(Y_{in}).$$

[1] Ralph E. Gomory, "Outline of an Algorithm for Integer Solutions to Linear Programs," *Bulletin of the American Mathematical Society*, 64 (September 1958): 275–278; and "An Algorithm for Integer Solutions to Linear Programs," *Princeton IBM Mathematical Research Report* (November 1958).

[2] Leon S. Lasdon, *Optimization Theory for Large Systems*, 1970.

3. Place x_v in the basis and add to the optimal tableau the new row of coefficients

$$-f(Y_{i1}),\ -f(Y_{i2}),\ \ldots,\ -f(Y_{in-1}),\ -f(Y_{in}).$$

4. Continue pivoting by the Dual Simplex Method until the tableau is feasible.
5. If all basic variables are not integral, then repeat the first four steps.

8.4 Examples

For our first example let us solve the following problem by Gomory's Algorithm and graph its solution.

Example 1. Find a pair of nonnegative integers x_1, x_2 that satisfy the constraints

$$-x_1 + 5x_2 \le 25$$
$$2x_1 + 1x_2 \le 24,$$

and $10x_2 - x_1$ is maximum. Let $M = 10x_2 - x_1$ or $M + x_1 - 10x_2 = 0$. Using x_3 and x_4 as the slack variables, the initial tableau is easily set up.

	1	2	
3	−1	5	25
4	2	1	24
	1	−10	0

The second tableau is found by the Simplex Method, pivoting at position (1, 2).

	1	3	
2	$-\frac{1}{5}$	$\frac{1}{5}$	5
4	$\frac{11}{5}$	$-\frac{1}{5}$	19
	−1	2	50

The third and optimal tableau is found by pivoting at (2, 1).

	4	3	
2	$\frac{1}{11}$	$\frac{2}{11}$	$6\frac{8}{11}$
1	$\frac{5}{11}$	$-\frac{1}{11}$	$8\frac{7}{11}$
	$\frac{5}{11}$	$\frac{21}{11}$	$58\frac{7}{11}$

The optimal solution for the original feasible region is

$$\begin{aligned} x_1 &= 8\tfrac{7}{11} \\ x_2 &= 6\tfrac{8}{11} \\ M &= 58\tfrac{7}{11}. \end{aligned}$$

Unfortunately x_1 and x_2 are not integers, so we apply Gomory's Algorithm. The basic variable with the larger fraction part is x_2. Therefore, we will form our cutting plane constraint from the first row of the optimal tableau as follows:

$$\begin{aligned} f(\tfrac{1}{11})x_4 + f(\tfrac{2}{11})x_3 &\geq f(6\tfrac{8}{11}) \\ \tfrac{1}{11}x_4 + \quad \tfrac{2}{11}x_3 &\geq \tfrac{8}{11}. \end{aligned}$$

Let x_5 be the new slack variable to be subtracted from the left side of the type II constraint. Then multiply this result by (-1) to form the new equation:

$$x_5 - \tfrac{1}{11}x_4 - \tfrac{2}{11}x_3 = -\tfrac{8}{11}.$$

The augmented optimal tableau appears as follows.

	4	3	
2	$\frac{1}{11}$	$\frac{2}{11}$	$6\frac{8}{11}$
1	$\frac{5}{11}$	$-\frac{1}{11}$	$8\frac{7}{11}$
5	$-\frac{1}{11}$	$-\frac{2}{11}$	$-\frac{8}{11}$
	$\frac{5}{11}$	$\frac{21}{11}$	$58\frac{7}{11}$

By the Dual Simplex Algorithm the next pivot is found to be at (3, 1). One pivoting iteration produces a feasible tableau in which all variables are integral.

	5	3	
2	1	0	6
1	5	−1	5
4	−11	2	8
	5	1	55

The best integral solution is

$$x_1 = 5$$
$$x_2 = 6$$
$$M = 55.$$

Since the feasible region has been reduced by the cutting plane, the new maximum is slightly less than the maximum of the original feasible region. The optimal integral vertex (5, 6) cannot be obtained by any rounding off process on the optimal vertex $(8\frac{7}{11}, 6\frac{8}{11})$. As a matter of fact the rounded off answer (9, 7) is not even feasible. Even if we try a nearby feasible point, such as (9, 6) where $M = 51$ or (8, 6) where $M = 52$, the objective function does not reach its maximum for integral coordinates.

A graphical analysis is helpful. We first draw the original feasible region of the given constraints, as shown in Figure 8–1. This region is the quadrilateral with vertices M, N, O, P.

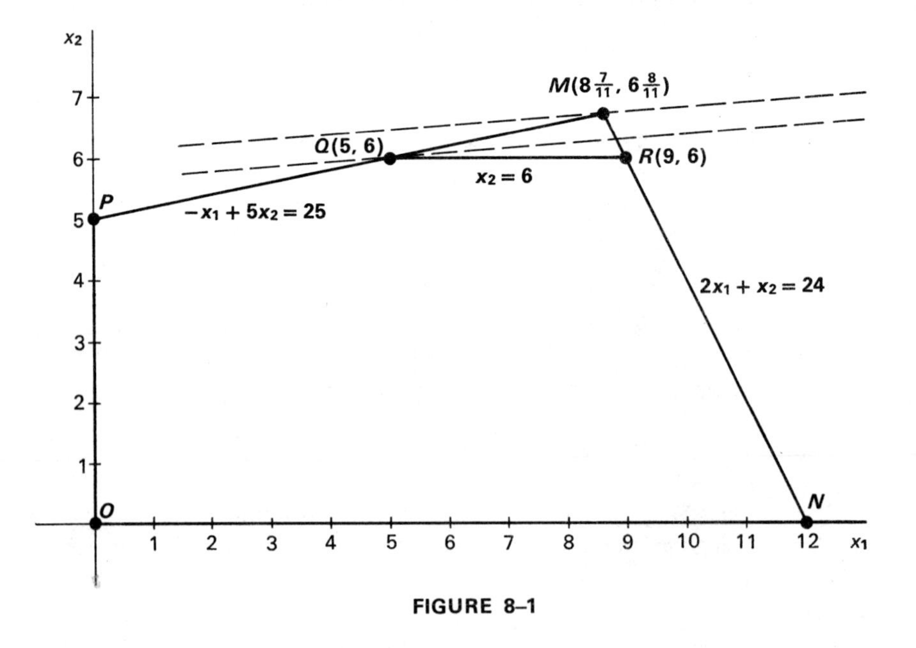

FIGURE 8–1

The objective function $-x_1 + 10x_2$ is a family of lines of slope $1/10$. Two members of this family are shown as dashed lines in Figure 8–1. The one through vertex M determines the original maximum of $58\frac{7}{11}$. If this line of slope $1/10$ is moved into the feasible region keeping its same slope, the first feasible point encountered with integral coordinates is (5, 6) as shown. The actual Gomory constraint is in the space of variables x_3 and x_4. The cut there determines a new vertex at $x_3 = 0$, $x_4 = 8$. These values of the slack variables in turn determine $x_1 = 5$ and $x_2 = 6$ in the given constraints.

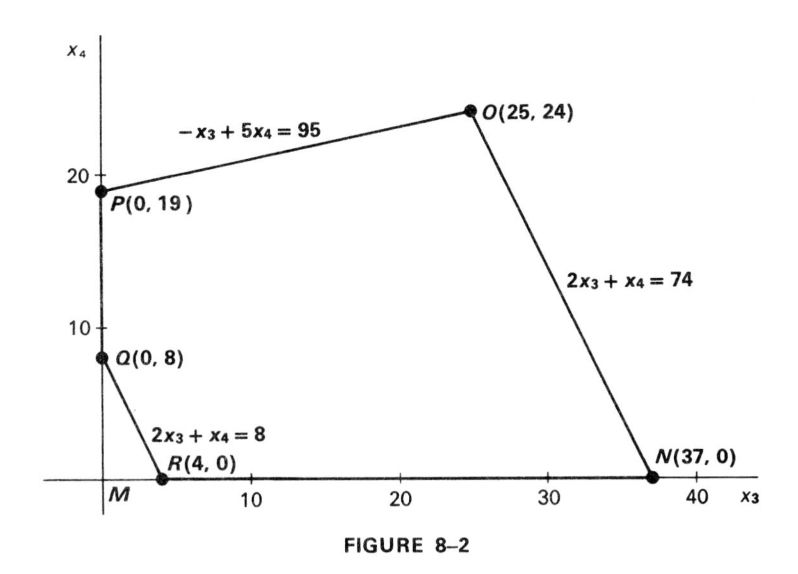

FIGURE 8–2

Figure 8–2 shows the Gomory cut as the line segment QR in the plane of variables x_3 and x_4. The remaining constraints in addition to $x_3 \geq 0$ and $x_4 \geq 0$ may be found from the optimal fractional tableau. After clearing fractions they are

$$\begin{aligned} 2x_3 + x_4 &\leq 74 \\ -x_3 + 5x_4 &\leq 95. \end{aligned}$$

In Figure 8–1 the cut in the plane of x_1 and x_2 that corresponds to the Gomory cut is also labeled QR. It may be determined by substituting the values of x_3 and x_4 from the original equations into the Gomory constraint as follows:

$$\begin{aligned} 2x_3 + x_4 &\geq 8 \\ 2(25 + x_1 - 5x_2) + (24 - 2x_1 - x_2) &\geq 8 \\ 74 - 11x_2 &\geq 8 \\ x_2 &\leq 6. \end{aligned}$$

The Gomory cut in Figure 8–2 removes the vertex at the origin and creates new vertices at (0, 8) and (4, 0). In our four dimensional solution space the sequence of vertices corresponding to the basic solutions of our tableaux is in the order

$$O(0, 0, 25, 24)$$
$$P(0, 5, 0, 19)$$
$$M(8\tfrac{7}{11}, 6\tfrac{8}{11}, 0, 0)$$
$$Q(5, 6, 0, 8).$$

The cut $x_2 \leq 6$ that we defined in Figure 8–1 did not remove any points with integral coordinates. The integral solution for the reduced region is therefore the same as the integral solution over the original feasible region. In a more complicated problem a number of cuts may have to be made to produce the integral answer. ●

Example 2. For another example of integer programming let us consider again Example 5, page 45. The production schedule for minimum cost required the making of a fractional number of bombs. In Chapter 3 we settled for the closest integers as approximate answers. Let us now find the best integer solution to the problem.

Solution. The original constraints and final tableau are reproduced below where x_1, x_2, and x_3 are the number of bombs to be produced of types *A*, *B* and *C* respectively.

$$\langle 1 \rangle \quad 3x_1 + x_2 + 6x_3 = 2000$$
$$\langle 2 \rangle \quad 2x_1 + 5x_2 + x_3 \geq 1000$$
$$\langle 3 \rangle \quad x_1 + 2x_2 + 4x_3 \leq 3000$$
$$1600x_1 + 3200x_2 + 2300x_3 = \text{minimum cost.}$$

The slack variables for constraints $\langle 2 \rangle$ and $\langle 3 \rangle$ were x_4 and x_7 respectively. The final tableau found in section 3.5 was

	2	4	
3	−1.444	.333	111.11
1	3.222	−.667	444.44
7	4.556	−.667	2111.11
	1366.7	300	−966,667

The previous integer solution was

$$x_1 = 444 \text{ bombs of type } A$$
$$x_2 = 0 \text{ bombs of type } B$$
$$x_3 = 111 \text{ bombs of type } C.$$

The solution is quite different after applying Gomory's Algorithm. We attempt to make x_1 integral by adding the new constraint

$$f(3.222)x_2 + f(-.667)x_4 \geq f(444.44)$$
$$.222x_2 + .333x_4 \geq .44.$$

Let x_8 be the new slack variable and then the equation to be added is

$$x_8 - .222x_2 - .333x_4 = -.44.$$

The following optimal tableau is then infeasible.

	2	4	
3	−1.444	.333	111.11
1	3.222	−.667	444.44
7	4.556	−.667	2111.11
8	−.222	−.333	−.44
	1366.7	300	−966,667

By the Dual Simplex Method a pivot is carried out at (4, 2) giving

	2	8	
3	−1.667	1	110.67
1	3.667	−2	445.33
7	5	−2	2112
4	.667	−3	1.33
	1166.67	900	−967,067

The solution is still not integral. Variable x_4 came into the basis at a nonintegral value and so we failed to make x_1 integral. Another cutting plane constraint is needed This time we work on x_3 to form

$$f(-1.667)x_2 + f(1)x_8 \geq f(110.67)$$
$$.333x_2 + 0 \geq .67.$$

The equation to be added with new slack variable x_9 is

$$x_9 - .333x_2 + 0 = -.67.$$

	2	8	
3	−1.667	1	110.67
1	3.667	−2	445.33
7	5	−2	2112
4	.667	−3	1.33
9	−.333	0	−.67
	1166.67	900	−967,067

After adding in x_9 the above tableau is pivoted at (5, 1) to obtain the final tableau.

	9	8	
3	−5	1	114
1	11	−2	438
7	15	−2	2102
4	2	−3	0
2	−3	0	2
	3500	900	—969,400

The final tableau is completely integral with optimal solution

$$x_1 = 438 \text{ bombs of type } A$$
$$x_2 = 2 \text{ bombs of type } B$$
$$x_3 = 114 \text{ bombs of type } C$$
$$m = -M = \$969{,}400.$$

All of the discrepancies in the Chapter 3 answer are now cleared up as the new solution satisfies each constraint precisely with slack $x_4 = 0$ and $x_7 = 2102$. Also the cost may be checked in the objective function:

$$\$1600\,(438) + \$3200\,(2) + \$2300\,(114) = \$700{,}800 + \$6{,}400 + \$262{,}000$$
$$= \$969{,}400.$$

Since the cutting planes have reduced the size of the feasible region, our cost for the integral solution is higher than the optimal cost for the original feasible region. However, we have the minimum cost for an integral solution that satisfies all the conditions in the problem. The rounded off solution was actually not in the feasible region. Finally, it is interesting to see that the integral solution even involves a different set of basic variables. Variable x_2 has come into the basis which could not have been predicted by merely looking at the fractional answers. ●

If a problem requires many cuts, the process may be speeded up by adding a number of cutting plane constraints at the same time. Nothing in the finiteness proof restricts the number of constraints that may be added simultaneously. In this case each new constraint is constructed from an appropriate row of the optimal tableau. Pivoting would then continue until the new tableau is feasible.

Problems Section 8.4

8–1. Use Gomory's Algorithm to solve the following problem for nonnegative integers x_1 and x_2.

$$x_1 + 3x_2 \leq 12$$
$$5x_1 + 8x_2 \leq 40,$$

where $x_1 + 2x_2 = M$, a maximum. Graph the feasible region for this problem and show a cutting plane through the integral solution.

8–2. Follow the directions to Problem 8–1 for solving and graphing the integer solution to the following problem.

$$x_1 + 2x_2 \leq 12$$
$$4x_1 + x_2 \leq 32,$$

where $2x_1 + 3x_2 = M$ is maximum. Can the integer solution be found by rounding off the optimal solution?

8–3. Solve for and graph the integer solution to

$$x_1 \geq 0, \; x_2 \geq 0$$
$$x_1 + 5x_2 \geq 25$$
$$2x_1 - 3x_2 \leq 8,$$

such that $x_1 + 10x_2 = m$, a minimum. On your graph of the feasible region plot the fractional solutions corresponding to each iteration of Gomory's Algorithm leading up to the integral solution.

8–4. For Problem 8–1 graph the Gomory cut in the plane of slack variables x_3 and x_4. Show how this cut produces a vertex that corresponds to the integer solution for x_1 and x_2.

8–5. Maximize $3x_1 + 6x_2 + 2x_3$ subject to the following constraints where x_i, $i = 1, 2, 3$, are nonnegative integers.

$$\begin{aligned} 3x_1 + 4x_2 + x_3 &\le 24 \\ x_1 + 3x_2 + 2x_3 &\le 12 \end{aligned}$$

8–6. Find nonnegative integers x_i, $i = 1, \ldots, 5$, satisfying the following constraints and yielding the maximum value of the objective function.

$$\begin{array}{llllll} x_1 & & & & & \le 15 \\ & x_2 & & - x_4 & & \le 0 \\ & & x_3 & & - x_5 & \le 0 \\ .3x_1 & & & + x_4 & & = 9 \\ .18x_1 + & .3x_2 & & & + x_5 & = 5.4 \\ .06x_1 + & .15x_2 + & .3x_3 & & & + 7 = M \end{array}$$

List the entire solution vector including the slack variables for the first three constraints. Are the slack variables also integral?

8–7. Does the following problem have a nonnegative integral solution? If so, find it. If not, explain why no integral solution exists and state the general solution.

$$\begin{array}{llllll} x_1 & & & & & \le 3 \\ & x_2 & & - x_4 & & \le 0 \\ & & x_3 & & - x_5 & \le 0 \\ .3x_1 & & & + x_4 & & = 1.8 \\ .18x_1 + & .3x_2 & & & + x_5 & = 1.08 \\ .06x_1 + & .15x_2 + & .3x_3 & & & + 7 = M, \text{ a maximum.} \end{array}$$

8–8. Find nonnegative integers x_i, $i = 1, \ldots, 5$, such that

$$\begin{aligned} 4x_1 + 3x_2 - 2x_3 + 2x_4 - x_5 &\le 12 \\ 2x_1 + 3x_2 + x_3 + 3x_4 + x_5 &\le 15 \\ 3x_1 + 2x_2 + x_3 + 2x_4 + 5x_5 &\le 20 \\ 2x_1 + 4x_2 + x_3 + 6x_4 + x_5 &\le 25 \\ x_3 &\le 3, \end{aligned}$$

where $3x_1 + 2x_2 + 3x_3 + 4x_4 + x_5 = M$ is a maximum.

8–9. Find nonnegative integers x_i, $i = 1, 2, \ldots, 5$, satisfying

$$\begin{aligned} x_1 + 2x_3 + 5x_5 &\le 5 \\ 2x_1 + x_2 + 4x_3 + x_4 + 2x_5 &\le 20 \\ x_1 + 3x_2 + x_3 + 2x_4 + 3x_5 &\le 33 \\ 3x_1 + 4x_2 + x_4 &\le 39, \end{aligned}$$

such that $3x_1 + 2x_2 + x_3 + 4x_4 + x_5 = M$, a maximum.

8–10. The Capital Cookie Company makes three types of cookies whose principal ingredients are flour, sugar, chocolate, and butter. The requirements for each type of cookie per pound and resources on hand are listed in the following table in pounds.

	flour	sugar	chocolate	butter
type 1	.5	.25	0	.25
type 2	.3	.3	.2	.2
type 3	.25	.25	.15	.35
resources	100	60	50	70

The profit per pound on each type of cookie is 25¢, 29¢, and 24¢ respectively. Type 1 is packed in 5 pound boxes, while type 2 is packed in 2 pound cartons and type 3 is put into 1 pound bags. Determine the best mix of packed cookies and the maximum profit. Which of the four resources would you increase to make the largest increase in profit?

8.5 All Integer Tableaux

If the original tableau has all integer entries, then there is a more recent technique that preserves the integer characteristic in each tableau of all subsequent iterations. This modified simplex algorithm is also due to Gomory.[3] The initial tableau is not first optimized as in the previous method. The idea is to add a cutting plane constraint before each iteration in such a way that the pivot element will be -1. Then a pivot by the Dual Simplex Method on -1 maintains all integers in the new tableau. The initial tableau must be set up as required by the Dual Simplex Method, that is, optimal but not feasible. Then a finite number of iterations with these new constraints leads to an optimal feasible all integer tableau.

The determination of the cutting hyperplane is more difficult than before. It involves the concept of the *greatest integer* in any rational number.

Definition 4. The **greatest integer** in any rational number r is the largest integer less than or equal to r.

[3] Ralph E. Gomory, "Research Report RC–189," IBM Research Center, Yorktown Heights, N. Y. (January 1960).

The greatest integer in r will be denoted by $[r]$. If r is negative, $[r]$ is necessarily negative. If $0 \le r < 1$, then $[r] = 0$. Some examples are

$$[-\tfrac{1}{3}] = -1$$
$$[-7.1] = -8$$
$$[\tfrac{9}{10}] = 0$$
$$[5.5] = 5$$
$$[-12] = -12.$$

The new constraint will be formed by dividing one of the rows of the current tableau by a rational number $\lambda > 1$, and then taking the greatest integer in each of the resulting numbers. The chosen row must have a negative entry in the constant column Y_{in}. The remaining entries in that row are the coefficients of the nonbasic variables. At least one of these coefficients must be negative or else the problem is not feasible. Let the chosen row be the ith row and let the nonbasic variables be $x_s, x_t, \ldots x_u$. Then the ith row corresponds to constraint

$$\langle 1 \rangle \qquad Y_{i1}x_s + Y_{i2}x_t + \cdots + Y_{in-1}x_u \le Y_{in}.$$

From ⟨**1**⟩ form the constraint

$$\langle 2 \rangle \qquad \left[\frac{Y_{i1}}{\lambda}\right]x_s + \left[\frac{Y_{i2}}{\lambda}\right]x_t + \cdots + \left[\frac{Y_{in-1}}{\lambda}\right]x_u \le \left[\frac{Y_{in}}{\lambda}\right].$$

For $\lambda > 1$ constraint ⟨**2**⟩ must be satisfied by any integer solution to the original constraint.[4] As long as λ is positive the sense of the inequality is preserved and the sum of the greatest integers can not be larger than the greatest integer in the sum. By choosing λ large enough, all

$$\left[\frac{Y_{ij}}{\lambda}\right] = -1 \quad \text{for} \quad Y_{ij} < 0.$$

This means that a pivot element of -1 is available in the jth column.

We also wish to keep λ as small as possible while gaining the pivot -1, since a smaller λ produces a greater change in the value of the objective function. This may be seen by substituting these numbers into our formula for the objective value after pivoting in the new row ⟨**2**⟩.

$$\overline{Y}_{mn} = Y_{mn} - (Y_{mq}\,Y_{pn}/Y_{pq})$$
$$= Y_{mn} - \left(Y_{mq}\left[\frac{Y_{in}}{\lambda}\right]\right)\Big/(-1)$$
$$\langle 3 \rangle \qquad = Y_{mn} + Y_{mq}\left[\frac{Y_{in}}{\lambda}\right]$$

[4] T. C. Hu, *Integer Programming and Network Flows*, 1969, pp. 247–248.

In a dual program the objective value decreases to its minimum. Because $Y_{mq} \geq 0$ and $Y_{in}/\lambda < 0$, equation ⟨**3**⟩ shows that the greatest possible decrease in Y_{mn} occurs for small λ.

After λ has been determined, constraint ⟨**2**⟩ is to be added to the current tableau with a new slack variable in the basis. The new row, call it the pth row, is chosen to be the pivotal row. Then

$$Y_{pj} = \left[\frac{Y_{ij}}{\lambda}\right], \quad j = 1, \ldots, n.$$

The pivotal column q is chosen according to the Dual Simplex Method by the minimum ratio $|\bar{\theta}_q|$ where

$$|\bar{\theta}_q| = \left|\frac{Y_{mq}}{Y_{pq}}\right| = \left|\frac{Y_{mq}}{[Y_{iq}/\lambda]}\right| = \left|\frac{Y_{mq}}{-1}\right| \leq \frac{Y_{mj}}{-[Y_{ij}/\lambda]}$$

for $Y_{ij} < 0$ and $j = 1, \ldots, n - 1$. Thus,

⟨**4**⟩
$$Y_{mq} \leq \frac{Y_{mj}}{-[Y_{ij}/\lambda]} \leq Y_{mj}$$

determines the pivotal column to be the one corresponding to the smallest entry in the objective row that is associated with a negative entry on the pivotal row. Let t_j be the largest integer such that

⟨**5**⟩
$$Y_{mq} \leq \frac{Y_{mj}}{t_j}$$

for $j = 1, \ldots, n - 1$ and corresponding $Y_{ij} < 0$.
From ⟨**4**⟩ and ⟨**5**⟩ and the definition of t_j, λ must satisfy

⟨**6**⟩
$$-[Y_{ij}/\lambda] \leq t_j, \qquad \text{for } Y_{ij} < 0.$$

Equation ⟨**6**⟩ says that an arbitrary integer, $-[Y_{ij}/\lambda]$ satisfying ⟨**4**⟩ must be no larger than the greatest possible integer t_j satisfying ⟨**5**⟩. The smallest λ that will satisfy ⟨**6**⟩ is

$$\lambda_j = -Y_{ij}/t_j \qquad \text{for each possible } j.$$

This is the λ that gives equality in ⟨**6**⟩. In order to satisfy ⟨**6**⟩ for all appropriate j, we pick the largest of the λ_j so the final value of λ is

⟨**7**⟩
$$\lambda = \max \lambda_j \qquad \text{for } Y_{ij} < 0, \ j = 1, \ldots, n - 1.$$

Notice that our determined value of λ preserves Y_{mq} as the $\bar{\theta}$ ratio of minimum absolute value since the value of the pivot will be -1. For the pivotal column q, $t_q = 1$ and $\lambda_q = -Y_{iq}$. Since $\lambda \geq \lambda_q$,

$$0 < \frac{-Y_{iq}}{\lambda} \leq \frac{-Y_{iq}}{\lambda_q} = 1$$

and thus the pivot $Y_{pq} = [Y_{iq}/\lambda] = -1$.

It may happen that $\max \lambda_j = 1$. This means that the generating row already has a pivot of -1 in the qth column. It is preferable to pivot immediately without adding a new constraint. Of course the integer characteristic is preserved. Cutting planes will be constructed only for $\lambda > 1$. The whole procedure is summarized in the following algorithm.

The All Integer Algorithm

1. Begin with an all integer tableau that is optimal but not feasible.
2. Select a row for generating the cutting plane by a random choice from those rows with a negative entry in the constant column. For a feasible problem this row, i, must have a negative entry, $Y_{ij} < 0$, for $j < n$.
3. Determine the pivotal column by the minimum entry in the objective row, Y_{mj}, that is associated with a negative Y_{ij}, $j < n$. Call this minimum entry Y_{mq}.
4. Define t_j to be the largest integers such that $Y_{mq} \leq Y_{mj}/t_j$ for $Y_{ij} < 0, j < n$. If $Y_{mq} = 0$, the t_j are undefined. In this case choose $\lambda = |Y_{iq}|$ and go to step 6.
5. Compute $\lambda_j = -Y_{ij}/t_j$, and let $\lambda = \max \lambda_j$.
6. If $\lambda = 1$ then pivot in the ith row at $Y_{iq} = -1$ without adding a constraint. Go to step 2 unless the tableau is feasible.
7. For $\lambda > 1$ construct the constraint in equation ⟨**2**⟩ by dividing Y_{ij} by λ and taking the greatest integer in each entry for $j = 1, \ldots, n$.
8. Add in the new row with a new slack variable, and pivot in this row at the qth column. The pivot entry will be -1.
9. Repeat steps 2 through 7 until the tableau is feasible or until no negative Y_{ij} can be found with a negative Y_{in}. In the latter case the problem has no integer solution.

The algorithm is explicit except in the choice of a generating row. Gomory has shown in his proof that the process will terminate in a finite number of steps, that either a cyclic or a random choice will work. Our previous choice of the row with the largest negative entry in the constant column is not covered under Gomory's proof although it seems to work in many cases.

To illustrate the whole process, let us again solve Example 5, page 45, that was handled by Gomory's Algorithm in section 8.4. This time we will start with the original constraints replacing the equality constraint with two inequalities. The problem may be stated as follows.

Example 3. Find nonnegative integers x_1, x_2, x_3 satisfying constraints

$$\begin{aligned} 3x_1 + x_2 + 6x_3 &\leq 2000 \\ 3x_1 + x_2 + 6x_3 &\geq 2000 \\ 2x_1 + 5x_2 + x_3 &\geq 1000 \\ x_1 + 2x_2 + 4x_3 &\leq 3000, \end{aligned}$$

where $1600x_1 + 3200x_2 + 2300x_3$ is a minimum.

As usual we let

$$\begin{aligned} m = -M &= 1600x_1 + 3200x_2 + 2300x_3 \\ M + 1600x_1 + 3200x_2 + 2300x_3 &= 0. \end{aligned}$$

The type II constraints are multiplied by -1 and the slack variables chosen to be x_4, x_5, x_6 and x_7. The initial tableau is

	1	2	3	
4	3	1	6	2000
5	−3	−1	−6	−2000
6	−2	−5	−1	−1000
7	1	2	4	3000
	1600	3200	2300	0

The initial tableau satisfies step 1 of the All Integer Algorithm. In step 2 we select the second row for generating the cutting hyperplane. All $Y_{2j} < 0$. By step 3 the pivotal column is column 1 and $Y_{mq} = Y_{51} = 1600$. In step 4 we define t_j by

$$\begin{aligned} 1600 &\leq 1600/t_1 \\ 1600 &\leq 3200/t_2 \\ 1600 &\leq 2300/t_3, \end{aligned}$$

so that $t_1 = 1$, $t_2 = 2$, $t_3 = 1$. Computing λ_j by step 5 we have

$$\begin{aligned} \lambda_1 &= -Y_{21}/t_1 = \tfrac{3}{1} = 3 \\ \lambda_2 &= -Y_{22}/t_2 = \tfrac{1}{2} \\ \lambda_3 &= -Y_{23}/t_3 = \tfrac{6}{1} = 6 \\ \lambda &= \max \lambda_j = 6. \end{aligned}$$

In step 7 we form constraint ⟨**2**⟩ as follows:

$$[-\tfrac{3}{6}]x_1 + [-\tfrac{1}{6}]x_2 + [-\tfrac{6}{6}]x_3 \leq [-\tfrac{2000}{6}]$$
$$-1x_1 \quad -1x_2 \quad -1x_3 \leq -334.$$

Choose x_8 to be the new slack variable, and then the cutting hyperplane to be added to the initial tableau is

$$x_8 - x_1 - x_2 - x_3 = -334.$$

The new row will be made row 5 so that the objective row is now row 6. The pivot position is at (5, 1) and of course the pivot value is -1. The result of this pivot is to bring x_1 into the basis as shown in the next tableau.

	8	2	3	
4	3	−2	3	998
5	−3	2	−3	−998
6	−2	−3	1	−332
7	1	1	3	2666
1	−1	1	1	334
	1600	1600	700	−534400

We select the second row to generate the new constraint and carry out the steps of our algorithm as follows.

$$Y_{mq} = Y_{63} = 700, \text{ so the pivotal column is column 3.}$$
$$700 \leq 1600/t_1,\ t_1 = 2$$
$$700 \leq 700/t_3,\ t_3 = 1$$
$$\lambda_1 = \tfrac{3}{2},\ \lambda_3 = \tfrac{3}{1}$$
$$\lambda = 3$$
$$[-\tfrac{3}{3}]x_8 + [\tfrac{2}{3}]x_2 + [-\tfrac{3}{3}]x_3 \leq [-\tfrac{998}{3}]$$
$$-1x_8 + 0 + (-1)x_3 \leq -333$$
$$x_9 - x_8 + 0 - x_3 = -333.$$

Slack x_9 is placed in the basis at row 6 and a pivot carried out at (6, 3) to obtain the next tableau.

	8	2	9	
4	0	−2	3	−1
5	0	2	−3	1
6	−3	−3	1	−665
7	−2	1	3	1667
1	−2	1	1	1
3	1	0	−1	333
	900	1600	700	−767500

This time the third row is picked to construct the cutting constraint as follows.

$$Y_{mq} = Y_{71} = 900, \text{ column 1 is pivotal.}$$
$$900 \leq 900/t_1, \; t_1 = 1$$
$$900 \leq 1600/t_2, \; t_2 = 1$$
$$\lambda_1 = 3, \; \lambda_2 = 3, \; \lambda = 3$$
$$[-\tfrac{3}{3}]x_8 + [-\tfrac{3}{3}]x_2 + [\tfrac{1}{3}]x_9 \leq [-\tfrac{665}{3}]$$
$$x_{10} - x_8 - x_2 + 0 = -222.$$

A pivot at (7, 1) in the new row produces the following tableau.

	10	2	9	
4	0	−2	3	−1
5	0	2	−3	1
6	−3	0	1	1
7	−2	3	3	2111
1	−2	3	1	445
3	1	−1	−1	111
8	−1	1	0	222
	900	700	700	−967300

The first row is the only row available for the next cutting constraint. Since only the second column has a negative entry in that row

$$\lambda = -Y_{12} = 2$$
$$0 + [-\tfrac{2}{2}]x_2 + [\tfrac{3}{2}]x_9 \leq [-\tfrac{1}{2}]$$
$$x_{11} + 0 - x_2 + x_9 = -1.$$

Adding this row and pivoting at (8, 2) we have

	10	11	9	
4	0	−2	1	1
5	0	2	(−1)	−1
6	−3	0	1	1
7	−2	3	6	2108
1	−2	3	4	442
3	1	−1	−2	112
8	−1	1	1	221
2	0	−1	−1	1
	900	700	1400	−968000

The second row has its only negative entry equal to −1, thus we already have a pivot of −1 available. $\lambda = 1$ and by step 6 we pivot at (2, 3) without adding a new constraint. This pivot produces the final tableau.

	10	11	5	
4	0	0	1	0
9	0	−2	−1	1
6	−3	2	1	0
7	−2	15	6	2102
1	−2	11	4	438
3	1	−5	−2	114
8	−1	3	1	220
2	0	−3	−1	2
	900	3500	1400	−969400

The solution agrees precisely with our previous integer solution.

$$x_1 = 438 \text{ bombs of type } A$$
$$x_2 = 2 \text{ bombs of type } B$$
$$x_3 = 114 \text{ bombs of type } C$$

where the minimum cost is $969,400. Slack variables x_4 and x_5 are necessarily zero since they arose from an equality. Slack variable $x_6 = 0$ corresponds to the previous slack variable x_4 and slack variable $x_7 = 2102$ as before. ●

The same number of pivots, five, was required by both techniques. The All Integer Algorithm has the advantage of avoiding the general solution and keeping all tableaux integral. Each algorithm has advantages on certain problems. However, it is not always possible to tell ahead of time which of the two will result in the fewer iterations on a given problem.

Problems Section 8.5

The following problems should be done by the All Integer Algorithm. As a check on your work they may also be solved by Gomory's Algorithm.

8–11. Find nonnegative integers x_1 and x_2 such that

$$\begin{aligned} x_1 + x_2 &\le 7 \\ 2x_1 - x_2 &\le -2 \\ x_1 + 6x_2 &\ge 6 \\ 2x_1 - 5x_2 &\le 5, \end{aligned}$$

where $x_1 + 14x_2 = m$ is a minimum. Graph the feasible region for this problem and show a cutting plane through the optimal integral point. What is the optimal vertex of the original feasible region?

8–12. Find nonnegative integers x_1 and x_2 such that

$$\begin{aligned} 4x_1 + 7x_2 &\le 28 \\ 2x_1 - x_2 &\ge -2 \\ 3x_1 + 10x_2 &\ge 15 \\ 2x_1 - 5x_2 &\le 5, \end{aligned}$$

where $2x_1 + 24x_2 = m$ is a minimum. Graph the feasible region and show a cutting plane through the optimal integral point. What is the optimal vertex of the original feasible region?

8–13. Work Problem 8–12 and answer the same questions if the objective function is changed to $m = 15x_1 + 24x_2$.

8–14. Find nonnegative integers x_i, $i = 1, 2, 3$, satisfying

$$\begin{aligned} x_1 - 2x_2 + 2x_3 &\le 5 \\ 2x_1 - x_2 - x_3 &\ge -13 \\ x_1 + 2x_2 + 3x_3 &\ge 7 \\ 3x_1 + x_2 + 7x_3 &\ge 14, \end{aligned}$$

such that $2x_1 + 4x_2 + 3x_3$ is a minimum. State the entire solution vector including all slack variables.

8–15. Find nonnegative integers x_i, $i = 1, 2, 3$ satisfying

$$\begin{aligned} 3x_1 + 2x_2 + x_3 &\geq 3 \\ 4x_1 + 6x_2 + 5x_3 &\geq 6 \\ x_1 + x_2 + 3x_3 &\geq 7, \end{aligned}$$

such that $5x_1 + 7x_2 + 10x_3$ is a minimum. State the entire solution vector including all slack variables.

8–16. Find nonnegative integers x_i, $i = 1, 2, 3$, satisfying

$$\begin{aligned} x_1 + 3x_2 + 2x_3 &\geq 5 \\ 2x_1 + x_2 - x_3 &\geq 4 \\ 4x_1 - x_2 + x_3 &\geq 7, \end{aligned}$$

such that $7x_1 + 3x_2 + 2x_3 = m$ is a minimum. Also give the values of all slack variables.

8–17. Find nonnegative integers x_i, $i = 1, 2, 3, 4$, satisfying

$$\begin{aligned} x_1 + 3x_2 + 2x_3 + 4x_4 &\geq 28 \\ 2x_1 + x_2 \qquad\quad + 3x_4 &\geq 16 \\ 2x_1 \qquad\quad + 5x_3 + x_4 &\leq 21 \\ 3x_1 + x_2 + 2x_3 + 4x_4 &\leq 28, \end{aligned}$$

such that $4x_1 + 3x_2 + 2x_3 + x_4 = m$ is a minimum. Give the values of all slack variables.

8–18. Carry out Problem 8–17 if the objective function is replaced by $m = 4x_1 + 3x_2 + 2x_3 + 5x_4$.

8–19. For the following problem find the complete solution vector in nonnegative integers.

$$\begin{aligned} 2x_1 + 3x_2 + 5x_3 + 2x_4 + 4x_5 &\geq 85 \\ 5x_1 + 2x_2 + 6x_3 + x_4 + 3x_5 &\geq 75 \\ 4x_1 + x_2 + 7x_3 + 3x_4 + 5x_5 &\geq 99 \\ 3x_1 + 4x_2 + 3x_3 + 5x_4 - 6x_5 &\geq 60, \end{aligned}$$

where $9x_1 + 5x_2 + 15x_3 + 2x_4 + 8x_5 = m$ is a minimum. Compare the integer solution with the general solution over the uncut feasible region.

8–20. For the following problem find the complete solution vector in nonnegative integers.

$$\begin{aligned} 3x_1 + 5x_2 + 8x_3 + 6x_4 &\geq 56 \\ 2x_1 + 7x_2 + 9x_3 - 4x_4 &\geq 10 \\ 8x_1 - 3x_2 + 5x_3 + 7x_4 &\geq 36 \\ 5x_1 + 6x_2 - 3x_3 + 4x_4 &\geq 54 \\ 7x_1 + 4x_2 + 6x_3 + 8x_4 &\geq 70, \end{aligned}$$

where $2x_1 + 4x_2 + 8x_3 + 5x_4 = m$ is a minimum. Compare the integer solution with the general solution over the uncut feasible region.

8–21. Solve Problem 8–3 by the All Integer Algorithm.

8–22. Find the integer solution to Problem 7–7, page 116. Is this solution equivalent to a rounding off of the general solution?

8–23. Find the integer solution to Problem 7–11, page 117. Can the integer solution be obtained by a rounding off of the general solution?

8–24. Attempt to solve Problem 8–1 using the All Integer Algorithm by the following technique. Dualize the problem in order to satisfy the initial conditions. Is the integer solution to the dual equivalent to the integer solution to the primal problem? Notice that the von Neumann Minimax Theorem only holds in the continuous case and not in the discrete case of integer solutions.

8–25. Dualize Problem 8–2 and find the minimum by the All Integer Algorithm. Is this minimum the same as the maximum to the primal problem found by Gomory's Algorithm?

8–26. The Boxcar Flying Service Company has a rush order to deliver supplies and parts to a neighboring country. They are packed in crates of two sizes. The volumes and weights of each size are given in the following table.

		cu. ft.	lbs.
sizes	1	60	400
	2	50	420

Two planes are available for today's run, Sue and Flo. Today's initial delivery must consist of at least 22 size 1 crates and at least 30 size 2 crates. Sue has a maximum load limit of 10,000 pounds of cargo and the limit for Flo is 12,000 pounds of cargo. The efficiency expert of Boxcar requires that Sue carry at least 1300 cubic feet of cargo on each flight while Flo must carry at least 1520 cubic feet of cargo on each flight. The total cost of handling and shipping these crates is \$3 per pound on Sue and \$3.50 per pound on Flo. What is the minimum cost of today's delivery and how should Boxcar load the crates on its two planes so as to achieve the minimum cost? Compare your integer solution to the solution found without using any cutting plane constraints.

8.6 Extensions of the All Integer Algorithm

In order to make the All Integer Algorithm applicable to primal programs where the initial tableau is feasible but not optimal, let us introduce an upper bound constraint. It is usually easy from the given constraints to determine an upper bound on the sum of the main variables. Of course one way to find a bound is to solve for the general solution and then use the next integer larger than the actual sum of these variables. The upper bound constraint is

$$\langle \mathbf{1} \rangle \qquad x_1 + x_2 + \cdots + x_{n-1} \leq B$$

where B is the predetermined integer bound. If constraint $\langle \mathbf{1} \rangle$ is added into the nonoptimal tableau, and then a pivot is performed in this row at the column with the most negative entry in the objective row, then the new tableau will be optimal but probably not feasible. Now the All Integer Algorithm may be applied until the tableau is feasible. All tableaux remain integral if the initial tableau is integral.

Example 4. Consider Example 1, page 128, that was solved by Gomory's Algorithm. The general solution was $(8\frac{7}{11}, 6\frac{8}{11})$ so an obvious upper bound on the sum is $B = 16$. Let the upper bound constraint be

$$x_1 + x_2 \leq 16.$$

If the new slack variable is x_5, then the initial tableau is as follows

	1	2	
3	−1	5	25
4	2	1	24
5	1	1	16
	1	−10	0

A pivot at (3, 2) produces the following optimal infeasible tableau.

	1	5	
3	−6	−5	−55
4	1	−1	8
2	1	1	16
	11	10	160

Using the All Integer Algorithm on row one, we find $\lambda = 6$, and the cutting plane is

$$-x_1 - x_5 \leq -10.$$

By adding in the new row with slack x_6 and pivoting at (4, 2), we have the next tableau.

	1	6	
3	−1	−5	−5
4	2	−1	18
2	0	1	6
5	1	−1	10
	1	10	60

This tableau has a natural pivot of −1 at (1, 1). Carrying out the pivoting gives the final tableau.

	3	6	
1	−1	5	5
4	2	−11	8
2	0	1	6
5	1	−6	5
	1	5	55

The integer solution is the same as that found previously.

$$M = 55 \quad \text{at} \quad (5, 6, 0, 8). \quad \bullet$$

The purpose of the All Integer Algorithm is to maintain the integer characteristic in all tableaux. However, there is nothing in the construction of the algorithm or in the proof of its finiteness that requires the constraint coefficients to be integers. In other words the algorithm may be applied to a tableau with real coefficients provided the objective row has all integer entries. Naturally the cutting plane has integer coefficients and the pivot will still be −1. Thus all variables coming into the basis will come in at integral levels. When the tableau is feasible the main variables will be integral in the solution. Slack variables may remain nonintegral.

Another remark of importance deals with the choice of a generating row for a cutting plane. Some cuts are stronger than others in the sense that fewer iterations are required to finish the job. In general if a natural pivot of -1 can be found in the tableau, it will be stronger than an arbitrary cut. This follows since a cut is computed from $\lambda > 1$, while $\lambda = 1$ produces a greater change in the objective value. As has been noted in step 6 of the All Integer Algorithm, a pivot should be taken in this case without adding a constraint. It is wise to look for a natural pivot of -1 while searching for a generating row. A good example of the improvement possible occurs in Problem 8–26, page 147. This problem may be completed in 5 iterations if the natural pivots are taken. If a cutting plane is added at every step, the problem mushrooms to over a dozen iterations.

A final observation is helpful in controlling the tableau size. Both the fractional and the all integer algorithms increase the tableau size with each cutting plane. After a number of iterations some of the cutting plane slack variables that were removed from the basis will be returned to the basis. A variable coming into the basis by pivoting necessarily comes in with a nonnegative value. The corresponding constraint for this reentering slack variable is thus satisfied by the current solution. This row may be immediately dropped from the tableau since it is redundant. The advantage is that after a certain point the tableau size will not increase as one row is dropped after each iteration. If a redundant row is not dropped, then at some later iteration it may again have a negative value in its last column. When this occurs the row may be used as a pivotal row or a cut generating row. However, there is little advantage to keeping the extra row. Further extensions will be considered in the problems.

Problems Section 8.6

8–27. Use an upper bound constraint along with the All Integer Algorithm to solve the following problem for nonnegative integers x_1 and x_2.

$$\begin{aligned} x_1 + 3x_2 &\le 12 \\ 5x_1 + 8x_2 &\le 40 \\ x_1 + 2x_2 &= M, \text{ a maximum.} \end{aligned}$$

8–28. Follow the directions to Problem 8–27 for finding the integer solution to

$$\begin{aligned} x_1 + 2x_2 &\le 12 \\ 4x_1 + x_2 &\le 32 \\ 2x_1 + 3x_2 &= M, \text{ a maximum.} \end{aligned}$$

Compare your result to the previous solution found by Gomory's Algorithm.

8–29. Find nonnegative integers x_1 and x_2 satisfying

$$\begin{aligned} 3x_1 - \quad x_2 &\geq -3 \\ x_1 + 18x_2 &\leq 109 \\ 23x_1 + \quad 6x_2 &\leq 161 \\ 4x_1 + 15x_2 &\geq 20, \end{aligned}$$

such that $x_1 + 3x_2$ is maximum. Compare your solution with the nonintegral solution.

8–30. Find nonnegative integers x_1 and x_2 such that $x_1 + 9x_2$ is maximum, subject to the same constraints given in Problem 8–29. Compare your solution with the nonintegral solution.

8–31. Find nonnegative integers x_1 and x_2 such that $7x_1 + 15x_2$ is minimum, subject to the same constraints given in Problem 8–29. Compare your solution with the nonintegral solution.

8–32. Find nonnegative integers x_1 and x_2 such that $4x_1 - x_2$ is minimum, subject to the same constraints given in 8–29. Compare your solution to the general solution over the uncut feasible region.

8–33. Solve Problem 8–6, page 136, by the All Integer Algorithm. Hint: Convert all coefficients into integers.

8–34. An intuitive Primal All Integer Algorithm may be developed as follows:

1. Start with a tableau that is feasible but not optimal and has all integer entries.
2. Choose the pivotal column q by the most negative number in the objective row.
3. Choose the source row i by the minimum θ ratio for all positive entries in the pivotal column q.
4. Let $\lambda = Y_{iq}$ for the chosen ith row and qth column. If $\lambda = 1$ then pivot on Y_{iq} and go to step 7.
5. If $\lambda > 1$ construct a cutting constraint by dividing the source row i by λ and then taking the greatest integer in each coefficient.
6. Add the cutting constraint with a new slack variable to the tableau and pivot in this row at the qth column.
7. Repeat steps 2 through 6 until the tableau is optimal or unbounded.

Try this algorithm on Example 4, page 148.

$$\begin{aligned} x_1 \geq 0, \; x_2 &\geq 0 \\ -x_1 + 5x_2 &\leq 25 \\ 2x_1 + \quad x_2 &\leq 24, \end{aligned}$$

where $-x_1 + 10x_2 = M$ is a maximum. Note that after redundant rows are dropped the solution is the same as before but requires more iterations. Unfortunately this algorithm does not admit a finiteness proof without further refinements.[5] However, it will work if the problem does not run into a cycle due to degeneracy.

8–35. Find the complete solution vector to the following problem where x_1, x_2, and x_3 are nonnegative integers. Assume the given coefficients are real numbers not necessarily rational.

$$\begin{aligned} 3.1416x_1 + 2.718x_2 + 1.414x_3 &\geq 12 \\ 1.732x_1 + 2.236x_2 + 2.718x_3 &\geq 14 \\ 2.111x_1 + 3.1416x_2 + 2.54x_3 &\geq 15, \end{aligned}$$

where $x_1 + x_2 + x_3 = m$ is a minimum.

8–36. Solve Problem 8–35 where the objective function is changed to $x_1 + 2x_2 + 3x_3 = m$.

8–37. It is possible to put all of the variables into the basis of the initial tableau. The nonbasic variables may be added to the basis by throwing in the identities $x_q - x_q = 0$ for each appropriate q. Try this on the example in Problem 8–34 and show the initial tableau is

	1	2	
1	−1	0	0
2	0	−1	0
3	−1	5	25
4	2	1	24
	1	−10	0

Carry out the Primal All Integer Algorithm by adding in each cutting constraint as row 5 and then dropping row 5 immediately after the pivot. The advantage is that the tableau remains the same size and the final tableau has x_i, $i = 1$ to 4, in the first 4 rows respectively.

[5] F. Glover, "A New Foundation for a Simplified Primal Integer Programming Algorithm," *Journal of the Operations Research Society of America*, 16 (4) (July–August 1968): 727–740.

9

PARAMETRIC PROGRAMMING

9.1 A Parameter in the Objective Function

One of the difficulties in setting up a linear program is that frequently some of the coefficients in the program are unknown. Costs and profits may vary rapidly. Quantities of raw material or inputs may be constantly changing. In general any coefficient in the initial tableau may be known only within some range. One way of solving many programs simultaneously and also trying to analyze the possible variations, is to add a parameter to the questionable coefficients. We will discuss two possible cases. One involves a parameter in the objective function and the other involves a parameter in the requirements column. The following problem uses a parameter to handle fluctuations in the objective function.

Example 1. A manufacturer wishes to determine the optimum production schedule for the coming month on three possible products for one of his assembly lines. Last month the three products sold for mark ups of \$80, \$40, and \$50 respectively. Next month the market is uncertain. It is not known how much higher or lower the mark up should be to meet the stiff competition. Still the manufacturer wants to know what the solutions should be corresponding to possible price variations and what choice of price he has within a given production schedule. Let x_1, x_2, and x_3 be respectively the numbers of the three products to be produced. The production constraints for next month on this assembly line are

$$\langle 1 \rangle \qquad 2x_2 + 3x_3 \leq 800$$

$$\langle 2 \rangle \qquad 2x_1 + x_2 + 4x_3 \leq 600$$

$$\langle 3 \rangle \qquad 4x_1 + x_2 + 2x_3 \leq 1000.$$

Solution. To solve the manufacturer's problem we assume that he wishes to maximize his gross profit determined by the mark up. His objective on last month's run was to maximize

$$\langle 4 \rangle \qquad 80x_1 + 40x_2 + 50x_3 .$$

We further assume that each of the three products is subject to the same price fluctuation so that each mark up will be changed by the same amount. By adding a parameter t to the coefficients of ⟨**4**⟩, we have the objective function for next month:

$$\langle 5 \rangle \qquad (80 + t)x_1 + (40 + t)x_2 + (50 + t)x_3 .$$

In order to keep track of the coefficients of t under pivoting, we add a row for t at the bottom of our tableau. The initial tableau appears as follows.

	1	2	3	
4	0	2	3	800
5	2	1	4	600
6	4	1	2	1000
	−80	−40	−50	0
t	−1	−1	−1	0

The initial tableau is optimal if $t \leq -80$ since the combined objective row will then have all positive entries. This makes economic sense. If the mark up decreases by $80 it will not be profitable to produce any of the three products. At the borderline case, $t = -80$, we have a possible pivot in the first column. This pivotal column is chosen by a zero entry in the combined objective row, even though such a pivot will not increase the objective value. The result of such a pivot is an alternate optimum. By checking the θ ratios this pivot is at (3, 1) giving the second tableau.

	6	2	3	
4	0	2	3	800
5	−.5	.5	3	100
1	.25	.25	.5	250
	20	−20	−10	20,000
t	.25	−.75	−.5	250

To see where the second tableau is optimal it is necessary to examine the inequalities

$$20 + .25t \geq 0$$
$$-20 - .75t \geq 0$$
$$-10 - .5\ t \geq 0.$$

The solution is

$$t \geq -80$$
$$t \leq -\tfrac{80}{3}$$
$$t \leq -20.$$

All three inequalities are satisfied if t is in the range

$$\langle \mathbf{6} \rangle \qquad -80 \leq t \leq -\tfrac{80}{3}.$$

For any value of t in the range of $\langle \mathbf{6} \rangle$ the second tableau is optimal. Choosing the end point $t = -\frac{80}{3}$, we find the next pivot in the second column to be at (2, 2). The third tableau is

	6	5	3	
4	2	−4	−9	400
2	−1	2	6	200
1	.5	−.5	−1	200
	0	40	110	24000
t	−.5	1.5	4	400

Once again we solve the appropriate inequalities for the range of t that makes the tableau optimal:

$$-.5t \geq 0 \quad \text{or} \quad t \leq 0$$
$$40 + 1.5t \geq 0 \quad \text{or} \quad t \geq -80/3$$
$$110 + 4t \geq 0 \quad \text{or} \quad t \geq -27.5.$$

The range satisfying all three inequalities for which the third tableau is optimal is

$$\langle \mathbf{7} \rangle \qquad -\tfrac{80}{3} \leq t \leq 0.$$

One more iteration is possible by choosing $t = 0$ and finding the pivot in the first column at (1, 1). This pivot produces the fourth tableau.

	4	5	3	
6	.5	−2	−4.5	200
2	.5	0	1	400
1	−.25	.5	1.5	100
	0	40	110	24000
t	.25	.5	1.5	500

The range of t for the fourth tableau to be optimal is all $t \geq 0$.

The four tableaux exhaust all real values for parameter t. The manufacturer may now read off his production schedule from the appropriate tableau for any price fluctuation. For example, if prices go down, he may lower last month's mark up by as much as $26.66 while using the production schedule of tableau three. His solution will be

$$x_1 = 200$$
$$x_2 = 200$$
$$x_3 = 0$$

with a profit of $24,000 + 400t$. A greater reduction in price would force him to use tableau two while an increase over last month's price would result in using tableau four. For any increase in price, $t > 0$, the solution will be

$$x_1 = 100$$
$$x_2 = 400$$
$$x_3 = 0$$

with a profit of $24,000 + 500t$. The manufacturer may set up his production schedule and then make last minute adjustments in price within the appropriate range while being assured of an optimal program. ●

9.2 A Parameter in the Requirements Column

The right-hand column of our initial tableau is called the requirements column. It contains the various inputs into the problem which may very well depend upon a parameter. We will treat such a parameter in a dual way to the method shown in Section 9.1. To keep track of the coefficients of the parameter, we add a column to the right side of each tableau. The initial tableau is first optimized and then succeeding pivots are chosen by the Dual Simplex Method. The following example will illustrate the procedure.

Example 2. Find the ranges of the parameter t for which $x_2 + x_3 - x_1$ is a maximum subject to the constraints

$$x_i \geq 0,\ i = 1, 2, 3$$

$$\langle 1 \rangle \qquad 2x_1 - 3x_2 + 4x_3 \leq 2$$

$$\langle 2 \rangle \qquad 2x_1 + x_2 - 2x_3 \geq t - 2$$

$$\langle 3 \rangle \qquad -x_1 + x_2 + 4x_3 \leq 4 - t.$$

Determine the solution for each of the possible ranges of parameter t.

Solution. The first step in the solution is to multiply constraint $\langle \mathbf{2} \rangle$ by -1 in order to get a system of type I constraints. Let x_4, x_5, and x_6 be the slack variables. This leads to a typical maximizing program for which the initial tableau is

	1	2	3		t
4	2	-3	4	2	0
5	-2	-1	2	2	-1
6	-1	1	4	4	-1
	1	-1	-1	0	0

The first phase of pivoting is to optimize the tableau by the Simplex Method so that the coefficients in the objective row are all positive. Of course, if there are any artificial variables in the basis, they should be eliminated. In our example a pivot is chosen in the second column at (3, 2). One pivot produces the following optimal tableau.

	1	6	3		t
4	-1	3	16	14	-3
5	-3	1	6	6	-2
2	-1	1	4	4	-1
	0	1	3	4	-1

While optimizing the initial tableau the parameter may be ignored except that the t column participates in the pivoting iterations. The optimal tableau is then examined for a range of t that will make it feasible. From the last two columns we have

$$\langle 4 \rangle \qquad \begin{aligned} 14 - 3t &\geq 0 \\ 6 - 2t &\geq 0 \\ 4 - t &\geq 0. \end{aligned}$$

The three inequalities in $\langle 4 \rangle$ are satisfied if $t \leq 3$. For $t \leq 3$ the current tableau is optimal and feasible with solution

$$\begin{aligned} x_1 &= 0 \\ x_2 &= 4 - t \\ x_3 &= 0 \end{aligned}$$

where $M = 4 - t$ is maximum. If $t = 3$, basic variable $x_5 = 0$, indicating a degenerate solution. Variable x_5 may be eliminated from the basis by pivoting at (2, 1) according to the Dual Simplex Algorithm. The result is the next tableau.

	5	6	3		t
4	−.333	2.667	14	12	−2.333
1	−.333	−.333	−2	−2	.667
2	−.333	.667	2	2	−.333
	0	1	3	4	−1

To find the feasibility range for this optimal tableau, we solve the inequalities

$$\langle 5 \rangle \qquad \begin{aligned} 12 - \tfrac{7}{3}t &\geq 0 \\ -2 + \tfrac{2}{3}t &\geq 0 \\ 2 - \tfrac{1}{3}t &\geq 0. \end{aligned}$$

Inequalities $\langle 5 \rangle$ are satisfied for t in the range

$$\langle 6 \rangle \qquad 3 \leq t \leq \tfrac{36}{7}.$$

For values of t in the range of $\langle 6 \rangle$, a solution to the problem is

$$\begin{aligned} x_1 &= -2 + \tfrac{2}{3}t \\ x_2 &= 2 - \tfrac{1}{3}t \\ x_3 &= 0 \\ M &= 4 - t. \end{aligned}$$

If $t = \frac{36}{7}$ basic variable $x_4 = 0$, and a pivot chosen by the Dual Simplex Algorithm occurs at (1, 1). This pivot produces

	4	6	3		t
5	-3	-8	-42	-36	7
1	-1	-3	-16	-14	3
2	-1	-2	-12	-10	2
	0	1	3	4	-1

The last tableau is optimal and feasible for all $t \geq \frac{36}{7}$. The problem solution for t in this range is

$$x_1 = -14 + 3t$$
$$x_2 = -10 + 2t$$
$$x_3 = 0$$
$$M = 4 - t. \quad \bullet$$

For another example of the use of parameters in linear programming, let us return to Example 4, page 111, and complete the *Sensitivity Analysis* suggested there.

Example 3. In Example 4 mentioned above we found the shadow prices of cotton and dacron to be $43\frac{1}{2}$ cents per ounce and 36 cents per ounce respectively. The question was raised as to how much additional cotton could be purchased without changing its shadow price. Likewise we wish to know how much additional dacron can be purchased without changing its shadow price. At first suppose we increase only the cotton and let t be the number of ounces of cotton purchased. Then suppose we increase only the dacron and let u be the number of ounces of dacron purchased. The problem is to solve

$$\langle 7 \rangle \qquad 2x_1 + 2x_2 \leq 240 + t$$

$$\langle 8 \rangle \qquad 3x_1 + 2x_2 \leq 260 + u$$

$$\langle 9 \rangle \qquad 3x_1 + x_2 \leq 220$$

$$\langle 10 \rangle \qquad 1.95x_1 + 1.59x_2 = M$$

for the ranges of t and u subject to the known shadow prices. The two parameters may be added in separate columns to the initial tableau as follows.

	1	2		t	u
3	2	2	240	1	0
4	3	2	260	0	1
5	3	1	220	0	0
	−1.95	−1.59	0	0	0

In three pivots the tableau is optimized, and the final tableau is given below.

	3	4		t	u
5	1.5	−2	60	1.5	−2
2	1.5	−1	100	1.5	−1
1	−1	1	20	−1	1
	.435	.36	198	.435	.36

If $u = 0$, the range of t for which this tableau is feasible can be found from

$$\langle \mathbf{11} \rangle \qquad \begin{aligned} 60 + 1.5t &\geq 0 \\ 100 + 1.5t &\geq 0 \\ 20 - \quad t &\geq 0. \end{aligned}$$

The solution to inequalities ⟨**11**⟩ is

$$\langle \mathbf{12} \rangle \qquad -40 \leq t \leq 20.$$

Range ⟨**12**⟩ shows that 20 additional ounces of cotton can be purchased while the shadow price remains $43\frac{1}{2}$ cents per ounce. The result agrees with our previous answer.

If $t = 0$, the range of u for feasibility is found from

$$\langle \mathbf{13} \rangle \qquad \begin{aligned} 60 - 2u &\geq 0 \\ 100 - \quad u &\geq 0 \\ 20 + \quad u &\geq 0 \end{aligned}$$

to be

$$\langle \mathbf{14} \rangle \qquad -20 \leq u \leq 30.$$

Range ⟨**14**⟩ shows that 30 additional ounces of dacron can be purchased while the shadow price remains 36 cents per ounce. It should be noted that the columns headed by t and u are identical to those headed by 3 and 4 in the final tableau. It was not necessary to add the columns t and u since they were each unit vectors. Recall that in condensing the extended tableau after a pivot, the column unit vector of the variable entering the basis was replaced with the column of coefficients of the variable just removed from the basis. Thus as variables x_1 and x_2 enter the basis, their corresponding columns of unit vectors t and u are replaced with the coefficients of x_3 and x_4 respectively. So it was necessary that column t correspond to x_3 and column u correspond to x_4. This insight grants us a use for the coefficients of a nonbasic variable in the final tableau. They may be used to determine how much a corresponding input can vary while this variable stays out of the basis. ●

If a variable represents the amount of slack raw material in one of the original problem constraints, then the fact that it is out of the basis in the final tableau means a number of important things. First of all that raw material is completely consumed by the optimal solution since its slack equals zero. The shadow price of that raw material is found at the bottom of its column in the objective row. Finally, the remaining coefficients in this column are the multiples of a parameter that added to the right-hand column and set greater than or equal to 0 will determine the interval over which the corresponding resource may vary in this solution. If this interval is exceeded the variable in question will be returned to the basis.

Example 4. One more question of interest from Example 4, page 111, is how much can the cotton and dacron be simultaneously increased while their shadow prices remain fixed. A limiting factor will be the amount of linen available. To answer this question we add a single parameter to both cotton and dacron in inequalities ⟨**7**⟩ and ⟨**8**⟩. The initial tableau is now

	1	2		t
3	2	2	240	1
4	3	2	260	1
5	3	1	220	0
	−1.95	−1.59	0	0

Again after three pivots we reach the optimal tableau

	3	4		t
5	1.5	-2	60	$-.5$
2	1.5	-1	100	.5
1	-1	1	20	0
	.435	.36	198	.795

The range of t for which this tableau is feasible is found from

$$60 - .5t \geq 0$$
$$100 + .5t \geq 0$$

to be

⟨**15**⟩ $$-200 \leq t \leq 120.$$

Range ⟨**15**⟩ indicates that cotton and dacron can simultaneously be increased by 120 ounces while their shadow prices remain the same. If we buy an additional 120 ounces of cotton and an additional 120 ounces of dacron the optimal production schedule is

$$x_1 = 20$$
$$x_2 = 160$$
$$M = 198 + .795(120) = \$293.40.$$

The three slack variables are all zero which means equality holds in the original constraints. In particular all of the slack linen is used up. ●

Our final tableau in parametric form solves a whole range of problems at the same time. Example 4 may be continued by the Dual Simplex Method to find new ranges of the parameter t and the corresponding optimal tableaux. Of course the shadow prices will change as soon as t increases past 120. It is left as an exercise to find the remaining ranges of t along with the new shadow prices.

Problems Chapter 9

For the first nine problems find the optimum value of the objective function for each of the possible ranges of parameter t that both optimizes the corresponding tableau and makes it feasible. All x_i are nonnegative.

9–1. $M = (5 + t)x_1 + (4 + t)x_2 + (8 + t)x_3$ such that
$5x_1 + 7x_2 + 2x_3 \leq 4000$
$2x_1 + 3x_2 + 4x_3 \leq 3000$
$4x_1 + x_2 + 2x_3 \leq 5000.$

9–2. $M = (7 + t)x_1 + (16 - t)x_2$ such that
$2x_1 + x_2 \leq 3$
$x_1 + 4x_2 \leq 4.$

9–3. $M = (8 + t)x_1 + (6 + t)x_2 + (4 + t)x_3$ such that
$5x_1 + 4x_2 + 2x_3 \leq 6000$
$2x_1 - 1.5x_2 + 5x_3 \leq 8000$
$4x_1 + x_2 + 2x_3 \leq 5000.$

9–4. $M = (5 + t)x_1 + (4 + t)x_2 + (t - 2)x_3$ such that
$8x_1 + 10x_2 + 5x_3 \leq 1000$
$5x_1 + 4x_2 + 12x_3 \leq 710$
$4x_1 + 8x_2 + 9x_3 \leq 745$
$8x_1 + 5x_2 + 4x_3 \leq 750.$
In particular what is the maximum and the solution vector if $t = -1$.

9–5. $m = 8x_1 + 3x_2 + 7x_3$ such that
$x_1 + 2x_2 - 3x_3 \leq 6$
$-2x_1 - x_2 + 2x_3 \leq 3 - t.$
In particular what is the minimum and the solution vector if $t = 5$? $t = 12$?

9–6. $m = 15x_1 + 6x_2 + 4x_3$ such that
$-x_1 + 2x_2 + 3x_3 \leq 5 + t$
$2x_1 + x_2 - x_3 \geq t - 2.$
From your parametric solution, find the minimum and the solution vector for $t = -6$.

9–7. $m = 5x_1 + 6x_2 + 4x_3$ such that
$x_1 + 2x_2 + 3x_3 \leq 5$
$-2x_1 + x_2 - x_3 \geq t - 2.$
Is there a solution for $t = 5$?

9–8. $M = x_1 + x_2 + x_3$ such that
$4x_1 - 5x_2 + x_3 \leq 50 + t$
$-5x_1 - 4x_3 \geq t - 20$
$2x_1 + 5x_2 \leq t + 60.$
From your parametric solution, find the maximum and the solution vector for $t = -50$.

9–9. $M = 3x_1 + 4x_2 - 2x_3$ such that
$2x_1 + 3x_2 - x_3 \leq t + 3$
$-x_1 - x_2 + 2x_3 \geq 2 - t$
$x_1 + x_2 + x_3 = 3.$

Are there any solutions for $t < -4$? For what value of t is the optimal solution $M = 0$? Does M continue to increase with t? Hint: Choose the first pivot so as to necessarily eliminate the artificial variable, and then continue by phase II pivoting until the tableau is optimal without regard to t.

9–10. In Example 3, page 159, find the additional solutions if cotton is increased beyond 20 ounces. What are the shadow prices of dacron and linen for such an increase in cotton?

9–11. In Example 3, page 159, find the additional solutions if dacron is increased beyond 30 ounces. What are the shadow prices of cotton and linen for such an increase in dacron?

9–12. In Example 4, page 161, suppose the dacron and cotton are simultaneously increased by more than 120 ounces. Find the possible new solutions along with the new shadow prices. In particular note the increase in the shadow price of linen as the demand for linen increases.

9–13. The Heavy Smelting Company has a plant that produces four alloys by blending together three metals. The resulting alloys, from different combinations of the base metals, achieve various degrees of the desired properties of strength, luster, hardness, and durability. The percentage of each of the base metals A, B, and C that occurs in the four alloys is given in the following table:

		metal		
		A	B	C
alloy	1	60	20	20
	2	75	25	0
	3	50	40	10
	4	80	0	20

There are 1000 tons of metal A, 500 tons of metal B, and 300 tons of metal C available for this month's production. Last month the four alloys sold for a gross profit of \$100, \$80, \$200, and \$75 per ton respectively. The selling prices at the end of this month are uncertain. Assume that the four alloys are subject to the same fluctuation in price so that the gross profit on each will be changed by the same amount. We wish to know how Heavy Smelting Company should plan its month's production to maximize gross profits. Determine all of the various possibilities due to possible price changes. In the most likely case of a relatively small change in selling price, what are the shadow prices of metal A, B, and C?

9–14. In Problem 9–13 the Heavy Smelting Company has a chance to buy metal B. Suppose that prices remain constant so that this month's gross profit is the same per ton as last month. How many more tons of metal B can be bought without changing the shadow price of B in the optimal solution? What is the maximum profit for each additional ton of B purchased up to this limit?

9–15. In Problem 9–13 the Heavy Smelting Company decides to buy more metal A. Suppose that prices remain constant so that this month's gross profit is the same per ton as last month. How many more tons of metal A can be bought without changing the basic variables in the optimal solution? What is the maximum profit for each additional ton of A purchased up to this limit?

9–16. In Problem 9–13 is there any price fluctuation for which it would be profitable to buy more metal C?

9–17. Suppose, in Problem 9–13, that resources A and B are simultaneously increased. By how many tons can both metals A and B be increased without changing the basic variables in the optimal solution and thus not changing the shadow prices? What is the new profit for each ton increase in both A and B up to this limit? Assume that the profit equation remains the same as last month.

9–18. Continue the analysis of Problem 9–17 by increasing both A and B beyond the previous limit. Find each of the possible new limits of increase along with the corresponding new shadow prices. In particular note the demand for metal C as its shadow price increases.

10

THE TRANSPORTATION PROBLEM

10.1 Definition of the Problem

The transportation problem in its direct form is the problem of minimizing the cost of shipping a commodity from a number of origins to various destinations. Suppose there are m origins and n destinations. At the ith origin there are $r_i > 0$ units of the product to be shipped and the number of units required by the jth destination is $s_j > 0$. We are initially given a cost matrix c_{ij}, $i = 1$ to m and $j = 1$ to n, where the entry or **cost coefficient** c_{ij} is the cost of shipping one unit from origin i to destination j. Since the origins correspond to rows and the destinations correspond to columns, the r_i are called **row requirements** and the s_j are called **column requirements**. These quantities may be displayed in tableau form as follows.

$$
\begin{array}{cc|cccc|cl}
 & & \multicolumn{4}{c}{\text{destinations}} & & \\
 & & 1 & 2 & \cdots & n & & \\
\hline
 & 1 & c_{11} & c_{12} & \cdots & c_{1n} & r_1 & \\
\text{origins} & 2 & c_{21} & c_{22} & \cdots & c_{2n} & r_2 & \text{row requirements} \\
 & \vdots & \vdots & \vdots & & \vdots & \vdots & \\
 & m & c_{m1} & c_{m2} & \cdots & c_{mn} & r_m & \\
\hline
 & & s_1 & s_2 & \cdots & s_n & & \\
 & & \multicolumn{4}{c}{\text{column requirements}} & &
\end{array}
$$

It will be assumed that the total supply equals the total demand, that is

$$\langle 1 \rangle \qquad \sum_{i=1}^{m} r_i = \sum_{j=1}^{n} s_j .$$

A solution to the problem is a matrix of nonnegative entries X_{ij} equal to the number of units shipped from origin i to destination j

$$\begin{array}{cccc} X_{11} & X_{12} & \cdots & X_{1n} \\ X_{21} & X_{22} & \cdots & X_{2n} \\ \vdots & \vdots & & \vdots \\ X_{m1} & X_{m2} & & X_{mn} \end{array}$$

such that

$$\langle 2 \rangle \qquad \begin{array}{l} \sum_{j=1}^{n} X_{ij} = r_i, \text{ for all } i \\ \sum_{i=1}^{m} X_{ij} = s_j, \text{ for all } j. \end{array}$$

The first set of equations in ⟨**2**⟩ says that the sum of each row in the solution matrix is equal to the corresponding row requirement. This means that all of the supply is shipped. The second set of equations in ⟨**2**⟩ says that the sum of each column in the solution matrix is equal to the corresponding column requirement. This means that demand is satisfied.

Each possible solution matrix has a cost given by

$$\langle 3 \rangle \qquad \overline{m} = \sum_{i=1}^{m} \sum_{j=1}^{n} c_{ij} X_{ij} .$$

The object of the transportation problem is to find a solution matrix for which the cost $\overline{m}$ is a minimum.

10.2 Northwest Corner Solutions

The following simple transportation problem will be used to illustrate the ideas as they are developed.

Example 1. A manufacturer has three factories at different locations that ship a certain product to three distributors around the country. He wishes to minimize his shipping costs. The shipping costs along with the current supply and demand are given in the tableau below.

		distributors 1	2	3		
factories	1	2	1	5	10	inventories
	2	7	4	3	25	
	3	6	2	4	20	
	orders	15	18	22		

Solution. The solution set is the 3×3 matrix

X_{11}	X_{12}	X_{13}
X_{21}	X_{22}	X_{23}
X_{31}	X_{32}	X_{33}

$X_{ij} \geq 0$

which must satisfy equations ⟨**2**⟩ and minimize equation ⟨**3**⟩. Notice that equation ⟨**1**⟩ is satisfied, the sum of the row requirements is 55 and the sum of the column requirements is 55. Equations ⟨**2**⟩ may be written out to get $m + n$ equations in mn unknowns.

$$\langle \mathbf{4} \rangle \quad \begin{aligned} X_{11} + X_{12} + X_{13} &= 10 \\ X_{21} + X_{22} + X_{23} &= 25 \\ X_{31} + X_{32} + X_{33} &= 20 \\ X_{11} + X_{21} + X_{31} &= 15 \\ X_{12} + X_{22} + X_{32} &= 18 \\ X_{13} + X_{23} + X_{33} &= 22 \end{aligned}$$

In this case we have 6 equations in 9 unknowns but one equation may be eliminated immediately by the fact that the sum of the first three is equal to the sum of the last three. There are at most $m + n - 1 = 5$ independent equations. As we might suspect and will eventually show, the optimal solution has at most $m + n - 1$ variables different from zero. In our example, dropping the last equation and setting $X_{12} = X_{13} = X_{31} = X_{32} = 0$ leads to the triangular system

$$
\langle 5 \rangle \qquad
\begin{array}{llllll}
X_{11} & & & & & = 10 \\
 & X_{22} & & & & = 18 \\
 & & X_{33} & & & = 20 \\
X_{11} & & & + X_{21} & & = 15 \\
 & X_{22} & & + X_{21} & + X_{23} & = 25.
\end{array}
$$

The system is triangular in the sense that every coefficient above the main diagonal is zero and thus every equation is immediately solvable from the preceeding ones. In this case and in general the equations ⟨**4**⟩ need to be reordered to exhibit the triangular feature in ⟨**5**⟩. However, the ordering of the equations in the system is immaterial. The important feature is the solvability. We agree to define a system of equations to be **triangular** provided the system can be solved one equation at a time using only the results of the previously solved equations. It is possible to arrange such a system in the form of ⟨**5**⟩. The solution matrix for equations ⟨**5**⟩ is

10	0	0
5	18	2
0	0	20

A quick way of arriving at this solution is the Northwest Corner Method. It may be described by beginning with a blank tableau that has the row requirements to the right and the column requirements underneath.

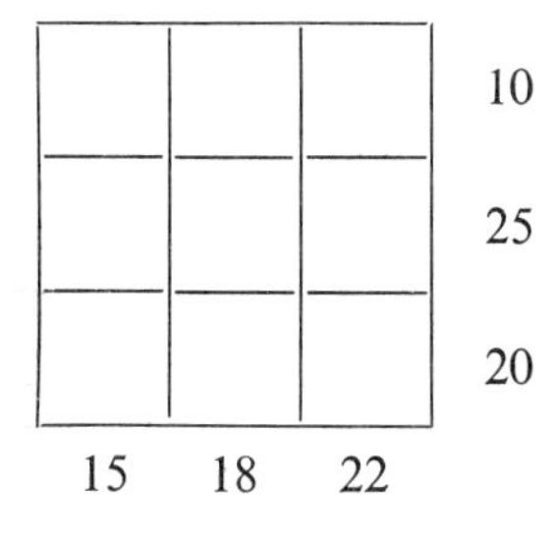

Starting at the northwest corner enter the smaller of the first row or first column requirement. Subtract off this entry from its corresponding row and column requirements.

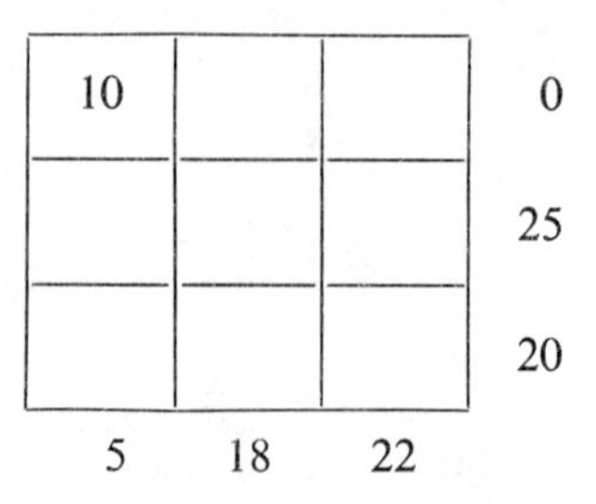

The first row is satisfied. Moving to the second row again enter the smaller of the appropriate requirements and subtract.

10			0
5			20
			20
0	18	22	

The first column is now satisfied. Move to the second column and always enter the smaller of the two requirements.

10			0
5	18		2
			20
0	0	22	

Each time move either right or down according to which requirement remains to be satisfied. The job is complete when all requirements are satisfied.

10			0
5	18	2	0
			20
0	0	20	

10			0
5	18	2	0
		20	0
0	0	0	

It is not necessary to start at the northwest corner. A similar process may be started at any matrix position. We could start at the southwest corner and then use a right-or-up rule to move from cell to cell. An alternate solution to our problem is displayed in the matrix below where the starting position was taken to be X_{32}. A left-or-up rule was used with the convention of associating the left column with the right column as if moving around a circle. Think of the matrix as wrapped around a circular cylinder so that the right and left edges coincide. It may also be necessary to associate the top and bottom edges of the matrix in the same way. Thus when position X_{21} was reached, the next move to the left appears in position X_{23}.

		10	10	0	
13		12	25	12	0
2	18		20	2	0
15	18	22			
13	0	10			
0		0			

The process leads to at most $m + n - 1$ nonzero entries since after the first choice there are $m - 1$ vertical moves and $n - 1$ horizontal moves for a total of $1 + m - 1 + n - 1 = m + n - 1$.

Definition 1. A nonzero X_{ij} is called a **solution variable**. The solution is **feasible** if all solution variables are positive. The solution is **basic** if the solution variables form a triangular system in equations ⟨**2**⟩.

Definition 2. If a solution contains less than $m + n - 1$ solution variables then the solution is **degenerate**.

The Northwest Corner Method necessarily leads to a basic feasible solution. Since the solution variables are determined one at a time the corresponding system of equations is necessarily triangular. The choice of the smaller requirement at each step insures that each solution variable will be positive so that the result is both feasible and basic.

If we consider starting at every matrix position and tracing out all possible paths, we generate all basic feasible solutions. Thus every basic feasible solution can be obtained by the Northwest Corner process. However, the solution found may be degenerate, and this situation will be handled in the next section.

10.3 Degeneracy

Theorem 10–1. A degenerate basic feasible solution exists if and only if some partial sum of the row requirements equals a partial sum of the column requirements.

Proof. Suppose $\sum_{i=1}^{p} r_i = \sum_{j=1}^{q} s_j$ for $p < m$ and $q < n$. Consider the Northwest Corner solution which yields at most $m + n - 1$ solution variables. From our hypothesis when we arrive at the pth row and qth column both a row and a column requirement will be satisfied simultaneously. This forces a diagonal step in the process, moving both a row and a column. Thus there is at least one less variable in the solution and the solution is degenerate.

Conversely, if we assume degeneracy and

$$\sum_{i=1}^{p} r_i \neq \sum_{j=1}^{q} s_j$$

for $p < m$ and $q < n$, then the Northwest Corner solution takes the usual $m + n - 1$ steps. This contradicts degeneracy and therefore equality holds for $p < m$ and $q < n$. If the partial sums do not consist of the first p rows and the first q columns then the matrix may be reordered until this condition is satisfied. Thus the theorem is true for any partial sum. ■

As an example consider a slight change in Example 1, page 167.

Example 2.

10			10	0	
4	21		25	21	0
		20	20	0	
14	21	20			
4	0	0			
0					

In this case the sum of the first two row requirements is 35 and the sum of the first two column requirements is 35. When we arrive at the (2, 2) position in the Northwest Corner solution, both the row and column requirements are satisfied. The next move must be a diagonal move to the (3, 3) position. We net only 4 solution variables instead of the required 5 for nondegeneracy. The degeneracy may be avoided by perturbing, that is, changing the requirements slightly. One way is by adding a small amount ε to each row requirement and then compensating by adding $m\varepsilon$ to one, say the last, column requirement.

In our case $\varepsilon = .1$ will do.

10.1			10.1
3.9	21	0.2	25.1
		20.1	20.1
14	21	20.3	

The result is 5 solution variables and the degeneracy is eliminated. ●

In a perturbed problem the correct solution to the original problem may be recovered from the final perturbed tableau by letting ε go to zero. If we define **equivalent transportation problems** to be a pair with the same cost matrix whose corresponding row and column requirements are arbitrarily close to one another, then equivalent problems have the same solution in the limit. The technique of perturbation leads to the following theorem.

Theorem 10–2. Every transportation problem with degenerate basic feasible solutions can be replaced with an equivalent problem in which degeneracy is impossible.

Since degeneracy may be eliminated we need to consider only non-degenerate problems in the rest of the chapter.

10.4 Finding Additional Basic Feasible Solutions

Once an initial solution has been found by the Northwest Corner Method, we wish to construct additional basic feasible solutions from the first and determine which is of least cost. The idea will be to introduce a new solution variable to replace one of the current solution variables. This may be done by first introducing a variable ϕ to a vacant cell in the current solution matrix. Then from this cell, trace out what is called the *plus-minus* path. Put a plus in the cell with ϕ and think of ϕ as a positive number added to that cell. To balance the row and column requirements, an amount ϕ must be subtracted from some solution variable in that row and also from some solution variable in that column. Continue to balance the row and column requirements by adding or subtracting ϕ from solution variables until the path returns to the initial cell. To illustrate the process consider the Northwest Corner solution of Example 1, page 167.

− 10		+
+ 5	18	− 2
		20

There are four vacant cells each of which determines a unique (+, −) path. If we wish to introduce variable X_{13} into the solution then the (+) at X_{13} is balanced in the first row by a (−) at X_{11}. Continuing, this requires a (+) at X_{21} and a (−) at X_{23} which balances the original (+) completing the path. The value of ϕ is determined by setting it equal to the smallest X_{ij} with a minus along the (+, −) path. In our case $\phi = 2$. Now adding or subtracting 2 according to the (+, −) path produces the new basic feasible solution.

8		2
7	18	
		20

Note that X_{22} or X_{33} could not be used along this (+, −) path because there would be no way to balance the change. If we introduce variable X_{31} into the Northwest Corner solution, then the (+, −) path begins with a (+) at cell (3, 1) to insure feasibility. To satisfy the third row requirement the same amount must be subtracted from 20 in cell (3, 3). Adding the same amount to cell (2, 3) balances the third column. The path is completed by subtracting this amount from cell (2, 1).

10		
− 5	18	+ 2
+		− 20

first B.F.S.

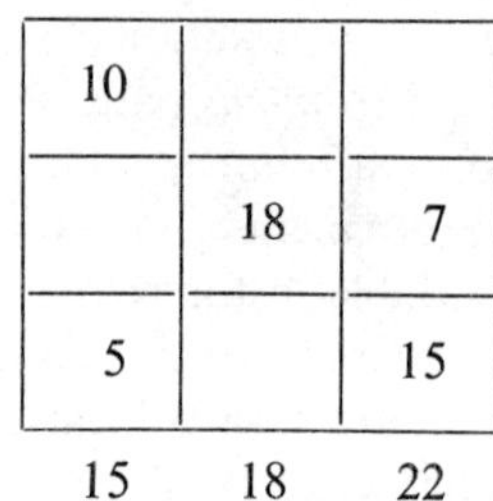

10		
	18	7
5		15
15	18	22

new B.F.S.

For values of ϕ strictly between 0 and 5 we have nonbasic solutions. At $\phi = 5$ we get the new basic solution shown above. For $\phi > 5$ the solutions would be infeasible.

If a transportation problem is degenerate it might be impossible to complete a $(+, -)$ path. Consider again Example 2, page 172.

$-$ 10		$+$	10
$+$ 4	$-$ 21		25
		$-$ 20	20
14	21	20	

There is no way to balance the second column or the third row. A $(+, -)$ path among the solution variables cannot be completed after starting at the vacant cell (1, 3).

On the other hand we will show that the nondegenerate case determines a unique path. For a nondegenerate transportation problem there can be no complete $(+, -)$ path that goes only through solution variables. If so, then the appropriate value of ϕ would knock out a solution variable leaving the next solution degenerate. Thus, every possible $(+, -)$ path must involve a plus in a vacant cell. If more than one vacant cell is assigned a $(+)$ then the subsequent solutions are nonbasic. Consider ϕ a nonnegative continuous variable assigned to a given vacant cell in our basic feasible solution matrix. Add or subtract ϕ along a $(+, -)$ path from this cell that involves only variables currently in the solution. Allow ϕ to increase continuously from zero. In the resulting solutions some solution variables will increase while others decrease proportionately. Nondegeneracy means that among the decreasing solution variables a unique one will reach zero first. The new feasible solution is therefore unique. If some other $(+, -)$ path through the solution variables were possible from this vacant cell, then a combination from the two would give a complete path among the solution variables contradicting the nondegeneracy. Thus only one path may be found.

We have verified the following theorem.

Theorem 10–3. From a nondegenerate transportation problem with a basic feasible solution, a new unique basic feasible solution may be derived starting at each vacant cell of the first solution.

We certainly don't want to introduce basic solutions at random. The question is which nonsolution variable should be brought in to reduce cost $\overline{m}$ the most. However, there is a more pressing question at the moment. If a transportation problem has an optimum solution, how do we know that this solution is basic? If the optimum is nonbasic then we are wasting time looking at basic solutions. The fact that a unique optimum must be basic is fundamental to the transportation algorithm.

10.5 The Fundamental Transportation Theorem

Theorem 10–4. If a nondegenerate transportation problem has a unique optimum solution, then that solution is a basic feasible solution.

Proof. Start with a nondegenerate transportation problem that has a unique optimum solution. A basic solution for this problem has exactly $m + n - 1$ solution variables. The total possible number of basic solutions is the binomial coefficient

$$\binom{mn}{m+n-1}.$$

Ignore those that are infeasible and consider all of the feasible basic solutions to be enumerated with their costs. Let

$$\overline{m} = \sum_{i=1}^{m} \sum_{j=1}^{n} c_{ij} X_{ij}$$

be the one of least cost. We will show that no nonbasic solution can have a smaller cost than $\overline{m}$. Since the optimum solution is unique only one basic solution can have the cost $\overline{m}$, and all nonbasic solutions will have to cost more. Thus the basic solution with cost $\overline{m}$ will be the optimum solution to the problem.

The demonstration that no nonbasic solution can have a cost smaller than $\overline{m}$ is carried out by induction on the number of solution variables. In the first case we show that this statement is true for $m + n$ solution variables. Introduce a new solution variable X_{ij} into a vacant cell of the basic solution of cost $\overline{m}$. Assume the cost m_1 of this nonbasic solution of $m + n$ variables satisfies

$$\langle 1 \rangle \qquad m_1 < \overline{m}.$$

Let E_{ij}, called the **entry cost** of X_{ij}, be the cost of adding in one unit along the $(+, -)$ path from X_{ij}. That is, $\phi = 1$ is added in along the $(+, -)$ path and E_{ij} stands for the change in cost from cost $\overline{m}$. Thus, for any positive value of ϕ the new cost is

$$\langle 2 \rangle \qquad m_1 = \overline{m} + \phi E_{ij}.$$

From equation ⟨**1**⟩, $E_{ij} < 0$. Now let ϕ increase until some solution variable is forced to zero. Then $\phi = \phi_{ij}$ is the value of the incoming variable X_{ij}, which coincides with the value of the unique variable removed from the solution. The cost m_2 of this new basic feasible solution is

⟨**3**⟩ $$m_2 = \overline{m} + \phi_{ij} E_{ij}$$

Since E_{ij} is negative, $m_2 < \overline{m}$. This contradicts the fact that $\overline{m}$ was the least cost among basic feasible solutions. The contradiction arose from the assumption in equation ⟨**1**⟩. Thus, no solution containing $m + n$ variables can have a cost less than $\overline{m}$.

For case two in the induction we will suppose that no solution with $m + n + k$ solution variables can have a cost less tham $\overline{m}$. The object is to show that this statement is true for $m + n + k + 1$ solution variables. Let m_k be the cost of an arbitrary solution with $m + n + k$ solution variables. Then

$$m_k \geq \overline{m}$$

by the induction hypothesis. Let E_{ij} be the entry cost of the $m + n + k + 1$ solution variable and let m_{k+1} be the cost of this solution. Then

⟨**4**⟩ $$m_{k+1} = m_k + \phi E_{ij}.$$

If $E_{ij} \geq 0$, then $m_{k+1} \geq m_k \geq \overline{m}$. If $E_{ij} < 0$, then allow ϕ to increase until some solution variable is forced to zero. For this value of ϕ, the solution reduces to the previous case of $m + n + k$ solution variables, and

⟨**5**⟩ $$m_k + \phi_{ij} E_{ij} \geq \overline{m}$$

by the induction hypothesis. For any

$$0 < \phi < \phi_{ij}$$

the inequality in ⟨**5**⟩ is strengthened and combining ⟨**4**⟩ with ⟨**5**⟩

$$m_{k+1} > \overline{m}.$$

Thus no solution with $m + n + k + 1$ solution variables has a smaller cost than $\overline{m}$. The induction gives the truth of this statement for all cases from $m + n$ to mn solution variables and we have completed the proof. ■

Theorem 10–4 does not answer the questions of existence or uniqueness. The existence question may be answered in the affirmative. In fact no feasible solution can have a cost less than the costs of all basic feasible solutions. Since there are only a finite number of basic feasible solutions, at least one must have the minimum cost. However, the solution may not be unique. If two basic solutions have the same minimum cost, say $m_2 = \overline{m}$, then equation ⟨**3**⟩,

$$m_2 = \overline{m} + \phi_{ij} E_{ij},$$

implies that the entry cost $E_{ij} = 0$. In this case X_{ij} may be brought into the solution at any value between zero and ϕ_{ij} without changing the cost. For $0 < X_{ij} < \phi_{ij}$ the corresponding solution will be optimal but *not* basic. Let us now return to the question of finding a basic solution of least cost.

10.6 Entry Costs

In order to find an additional basic feasible solution of less cost, it is necessary to compute the entry costs E_{ij} of the nonsolution variables. The total cost $\overline{m}$ of the new solution may be found from the cost m of the old solution by formula ⟨**3**⟩ in Section 10.5:

$$\overline{m} = m + \phi_{ij} E_{ij}.$$

It is clear from this formula that to reduce cost the E_{ij} must be negative. We will therefore consider only negative entry costs and use the so called **Rule of Steepest Descent**. Steepest descent means to choose the variable X_{ij} which has the most negative entry cost to enter the next solution. If all entry costs are known it is then obvious how to proceed.

To easily compute the E_{ij} we define a new set of variables U_i and V_j by

$$U_i + V_j = c_{ij}, \qquad \langle \mathbf{6} \rangle$$

for those ij coresponding to a solution variable. Let us return to the Northwest Corner solution of Example 1, page 167, and write out equations ⟨**6**⟩.

solution matrix

10			10
5	18	2	25
		20	20
15	18	22	

cost matrix

2	1	5
7	4	3
6	2	4

$$U_1 + V_1 = 2 = c_{11}$$
$$U_2 + V_1 = 7 = c_{21}$$
$$U_2 + V_2 = 4 = c_{22}$$
$$U_2 + V_3 = 3 = c_{23}$$
$$U_3 + V_3 = 4 = c_{33}.$$

We have 5 equations in 6 unknowns. Assuming that a transportation problem is nondegenerate, system **⟨6⟩** will lead to $m + n - 1$ equations in $m + n$ unknowns. Since one unknown may be assigned an arbitrary value, let us set $V_1 = 0$. The remaining system is triangular and is immediately solved one equation at a time. The solution is

⟨7⟩
$$\begin{aligned} U_1 &= 2 \qquad & V_1 &= 0 \\ U_2 &= 7 \qquad & V_2 &= -3 \\ U_3 &= 8 \qquad & V_3 &= -4. \end{aligned}$$

The entry costs for nonsolution variables are listed below in equations **⟨8⟩**. Each formula is found by adding or subtracting the costs along the $(+, -)$ path from that variable.

⟨8⟩
$$\begin{aligned} E_{12} &= c_{12} - c_{22} + c_{21} - c_{11} \\ E_{13} &= c_{13} - c_{23} + c_{21} - c_{11} \\ E_{31} &= c_{31} - c_{33} + c_{23} - c_{21} \\ E_{32} &= c_{32} - c_{33} + c_{23} - c_{22} \end{aligned}$$

Formulas **⟨8⟩** give the change in cost for $\phi = 1$ unit in equations **⟨2⟩**. We now substitute equations **⟨6⟩** into equations **⟨8⟩** and simplify.

⟨9⟩
$$\begin{aligned} E_{12} &= c_{12} - U_2 - V_2 + U_2 + V_1 - U_1 - V_1 \\ &= c_{12} - (U_1 + V_2) \\ E_{13} &= c_{13} - U_2 - V_3 + U_2 + V_1 - U_1 - V_1 \\ &= c_{13} - (U_1 + V_3) \\ E_{31} &= c_{31} - U_3 - V_3 + U_2 + V_3 - U_2 - V_1 \\ &= c_{31} - (U_3 + V_1) \\ E_{32} &= c_{32} - U_3 - V_3 + U_2 + V_3 - U_2 - V_2 \\ &= c_{32} - (U_3 + V_2). \end{aligned}$$

In general the U's and V's with common subscripts will cancel leaving only the U of the row entered and the V of the column entered by E_{ij}. Thus the formulas for computing entry costs are

⟨10⟩
$$E_{ij} = c_{ij} - (U_i + V_j).$$

In order to write ⟨**10**⟩ as a matrix equation, we define

$$\langle \mathbf{11} \rangle \qquad W_{ij} = U_i + V_j \quad \text{for} \quad i = 1 \text{ to } m \quad \text{and} \quad j = 1 \text{ to } n.$$

The W_{ij} matrix is independent of the value arbitrarily assigned to one of the U's or V's in order to solve system ⟨**6**⟩. The entry cost for each ij is then the cost matrix minus the W_{ij} matrix. For ij corresponding to a variable currently in the solution, $W_{ij} = c_{ij}$, and thus $E_{ij} = 0$ for all solution variables.

In the E_{ij} matrix we need to write only the E_{ij} of nonsolution variables. For our example the result is

c_{ij}

2	1	5
7	4	3
6	2	4

$-$ W_{ij}

2	−1	−2
7	4	3
8	5	4

$=$ E_{ij}

	2	7
−2	−3	

Using the Rule of Steepest Descent, X_{32}, of entry cost -3, should be brought into the solution next to form an improved basic feasible solution.

10.7 The Transportation Algorithm

We are now ready to formally state the Transportation Algorithm as an iterative procedure. For a nondegenerate transportation problem it will produce an optimal solution in a finite number of steps. If degeneracy occurs the problem should be perturbed by the method shown in section 10.3 before continuing with the algorithm.

The Transportation Algorithm (Minimum)

1. Start with an initial basic feasible solution which may be found by the Northwest Corner Method.
2. Construct the W_{ij} matrix as follows:
   (a) $W_{ij} = c_{ij}$ for ij corresponding to a solution variable.
   (b) Let $V_1 = 0$, and define sets U_i, $i = 1, \ldots, m$, and V_j, $j = 1, \ldots, n$, by $U_i + V_j = c_{ij}$ for those ij corresponding to a solution variable.
   (c) Then the remaining $W_{ij} = U_i + V_j$.

3. Find the entry costs E_{ij} for those variables out of the solution by $E_{ij} = c_{ij} - W_{ij}$. If all $E_{ij} \geq 0$ the solution is optimal.
4. Pick the variable X_{ij} with the most negative entry cost E_{ij} to enter the solution. In the solution matrix trace out the $(+, -)$ path from that variable. Set ϕ equal to the smallest X_{ij} with a $(-)$ in the $(+, -)$ path.
5. Compute a new basic feasible solution by adding or subtracting ϕ along the $(+, -)$ path.
6. Repeat steps 2 through 5 until some solution is optimal, that is, until all entry costs are nonnegative.
7. Note whether $E_{ij} = 0$ for a nonsolution variable in the final tableau. If so, there exist alternate optima that may be found by bringing the variable of zero entry cost into the solution.

If it is desired to find a maximum cost from the algorithm, choose the X_{ij} with the largest positive entry cost to enter the solution. In this case the iterations will terminate when all entry costs are nonpositive. Because there are only a finite number of basic solutions either the minimizing or the maximizing problem must be terminated in a finite number of steps.

To complete Example 1, page 167, let us combine all of the calculations into a single matrix. We divide each cell of the solution matrix into three parts as follows.

W_{ij}	E_{ij}
X_{ij}	

In addition the values of the solution variables will be circled to make the solution stand out. The V_j will be placed across the top of the tableau and the U_i will be placed in a column to the left. Following the steps of the algorithm in order, we arrive at the initial tableau given below:

cost matrix

2	1	5
7	4	3
6	2	4

solution matrix

	0	-3	-4	
2	2 (10)	-1 2	-2 7	10
7	7 (5)	4 (18)$-$	3 (2)+	25
8	8 -2	5 -3 +	4 (20)$-$	20
	15	18	22	

The initial tableau contains the Northwest Corner solution whose cost may be found directly by

$$\begin{aligned} m &= (2)10 + (7)5 + (4)18 + (3)2 + (4)20 \\ &= 213. \end{aligned}$$

As before X_{32} at entry cost -3 should come into the solution. The $(+, -)$ path from X_{32} has been indicated in the initial tableau. We set $\phi = 18$, the smallest solution variable with a $(-)$ along this path. Adding or subtracting $\phi = 18$ will eliminate X_{22} from the solution and give the next solution.

	0	-6	-4
2	2 (10)	-4 5	-2 7
7	7 (5)$-$	1 3	3 (20)+
8	8 -2 +	2 (18)	4 (2)$-$

The cost of the second solution may be found from

$$\begin{aligned} \overline{m} &= m + \phi_{32} E_{32} \\ &= 213 + 18(-3) = 159. \end{aligned}$$

Variable X_{31} should come into the solution next since it is the only nonsolution variable with a negative entry cost. The $(+, -)$ path from X_{31} shows that $\phi_{31} = 2$. Adding or subtracting 2 along the path will eliminate variable X_{33} and give the final tableau.

	0	−4	−4
2	2 (10)	−2 3	−2 7
7	7 (3)	3 1	3 (22)
6	6 (2)	2 (18)	2 2

Since all four entry costs of nonsolution variables are positive, the solution is optimal. The cost of the optimal solution may be found directly or from

$$\overline{m} = 159 + 2(-2) = 155.$$

Thus the minimum possible cost subject to the original conditions of Example 1, page 167, is 155. ●

10.8 Multiple Solutions

The following transportation problem will illustrate the possibility of multiple solutions.

Example 3.

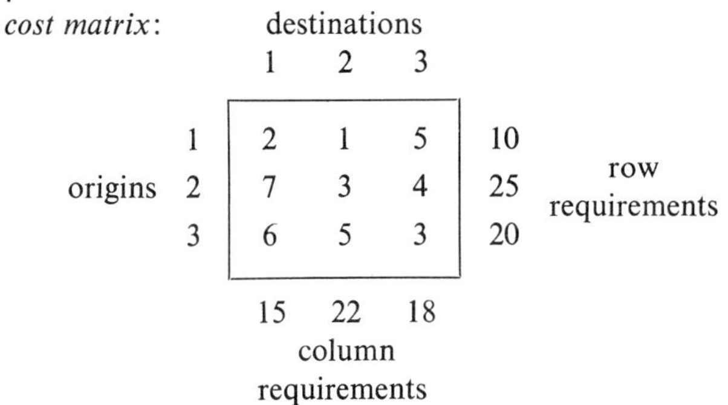

cost matrix: destinations

origins	1	2	3	row requirements
1	2	1	5	10
2	7	3	4	25
3	6	5	3	20
column requirements	15	22	18	

solution
matrix:

	0	−4	−6	
2	2 ⑩	−2 3	−4 9	10
7	7 ⑤−	3 ⑳+	1 3	25
9	9 −3 +	5 ②−	3 (18)	20
	15	22	18	

Variable X_{31} should come into the solution. The $(+, -)$ path determines $\phi = 2$ which eliminates X_{32} from the solution giving

	0	−4	−3
2	2 ⑩	−2 3	−1 6
7	7 ③−	3 (22)	4 0 +
6	6 ②+	2 3	3 (18)−

The cost of this solution from the tableau is

$$m = 20 + 21 + 12 + 66 + 54 = 173.$$

Since the entry costs are nonnegative the solution is optimal and 173 is the minimum cost. However, the zero entry cost for variable X_{23} indicates that this variable may be brought into the solution without changing the cost. Let us introduce X_{23}, choosing $\phi = 3$ which eliminates solution variable X_{21}. The result is an alternate optimum basic solution:

	0	-4	-3
2	2 10	-2 3	-1 6
7	7 0	3 22	4 3
6	6 5	2 3	3 15

The cost of this alternate optimum may be checked to be

$$m = 20 + 30 + 66 + 12 + 45 = 173.$$

The entry costs are seen to be the same as in the previous tableau.

Suppose instead of $\phi = 3$ we choose $0 < \phi < 3$ in the next to last tableau above. Then the solution obtained is nonbasic but still optimal. For $\phi = 2$ the nonbasic optimal solution is given in the next tableau.

2 10		
7 1	3 22	4 2
6 4		3 16

The cost of this nonbasic solution is also

$$m = 20 + 7 + 24 + 66 + 8 + 48 = 173. \quad \bullet$$

We see from this example that the transportation algorithm will necessarily give us an optimal solution but that the optimal solution may not be unique.

It is possible to have a transportation problem that has both degenerate and alternate optima. A degenerate problem might have $m + n - 2$ solution variables in a basic solution instead of the required $m + n - 1$

solution variables for a nondegenerate basic solution. Suppose a problem has two optimal basic solutions of $m + n - 2$ variables. Then nonbasic solutions of $m + n - 1$ solution variables may be formed by choosing ϕ between the two critical values corresponding to the two basic solutions, namely:

$$0 < \phi < \phi_{ij}.$$

The fact that a nonbasic solution might have the right number of variables for a basic solution forced us to define basic in terms of the solution to a triangular system instead of by a count of solution variables. If a nonbasic solution has $m + n - 1$ solution variables as described above then it cannot be found from a triangular system constructed out of the original equations.

10.9 Variations

There are a number of possible variations to the transportation problem that widen its applicability. If the context of a problem requires that a certain variable be excluded from the solution then this variable should be assigned an arbitrarily high cost. By setting the cost high enough any variable can be forced out of the solution. On the other hand if the problem requires that a certain variable must appear in the solution, then we force this variable into the solution by assigning it a small cost. Provided the cost is small enough our algorithm will bring the variable into the solution. Then the correct cost of the optimal solution is found by returning this variable to its original cost.

The condition that the sum of the row requirements be equal to the sum of the column requirements may be relaxed. If these sums are not equal we will create a dummy row or a dummy column to take up the slack. A dummy row means that the demand exceeds the supply so that some of the column requirements are not satisfied. A dummy column means that the supply exceeds the demand, so that some items will not be shipped. The cost coefficients of the dummy variables will be taken to be zero so that they will not change the objective function.

It has already been mentioned that the Transportation Algorithm may be used to find a maximum. Instead of changing the choice of incoming variable an easier way to maximize is to change the sign of all cost coefficients. In this way the same algorithm is used in both cases just as it was in linear programming. The principle involved is that maximizing the negative of the objective function is equivalent to minimizing the objective function. The Transportation Algorithm minimizes the objective function. If this function is replaced with its negative, then the same algorithm will maximize the objective function. Thus a single computer routine will handle both cases. The Transportation Algorithm is quite difficult to program although such programs do exist.[1] The computer language used is usually that of ALGOL. However, if the problem is not too large we may solve it by the Simplex Method and use our previously developed automatic routine in FORTRAN. The disadvantages of the Simplex Method are that it is quite inefficient compared to the Transportation Algorithm and requires a much larger tableau to work with.

10.10 Programming the Transportation Problem

In order to use the automatic pivoting routines developed in Chapter 4, we will now solve the transportation problem by the Simplex Method. In comparison, the advantages of the Transportation Algorithm should become apparent.

Example 4. Let us return to Example 1, page 167, in which the restraining equations were

$$\langle 1 \rangle \qquad \begin{aligned} X_{11} + X_{12} + X_{13} &= 10 \\ X_{21} + X_{22} + X_{23} &= 25 \\ X_{31} + X_{32} + X_{33} &= 20 \\ X_{11} + X_{21} + X_{31} &= 15 \\ X_{12} + X_{22} + X_{32} &= 18 \\ X_{13} + X_{23} + X_{33} &= 22 \end{aligned}$$

$$\langle 2 \rangle \qquad X_{ij} \geq 0 \quad \text{for} \quad i = 1, 2, 3 \quad \text{and} \quad j = 1, 2, 3.$$

Only $m + n - 1 = 5$ of equations $\langle 1 \rangle$ are independent so the last equation in that set will be ignored. The objective function is the cost equation

$$\langle 3 \rangle \quad \bar{m} = \sum_{i=1}^{m} \sum_{j=1}^{n} c_{ij} X_{ij}$$
$$= 2X_{11} + 1X_{12} + 5X_{13} + 7X_{21} + 4X_{22} + 3X_{23} + 6X_{31} + 2X_{32} + 4X_{33}.$$

[1] G. Bayer, "The Transportation Problem," *Communications of the Association of Computing Machinery* (December 1966): 869, Alg. 293.

Constraints ⟨**1**⟩ and ⟨**2**⟩ along with objective ⟨**3**⟩ constitute a standard linear program where the objective function is to be minimized. An artificial variable should be introduced for each equality constraint. For convenience let us rename our variables as follows:

$$x_1 = X_{11} \qquad x_4 = X_{21} \qquad x_7 = X_{31}$$
$$x_2 = X_{12} \qquad x_5 = X_{22} \qquad x_8 = X_{32}$$
$$x_3 = X_{13} \qquad x_6 = X_{23} \qquad x_9 = X_{33}.$$

The five artificial variables for the first five equations in ⟨**1**⟩ will be x_{-10}, x_{-11}, x_{-12}, x_{-13}, and x_{-14}. Each artificial variable is given an arbitrarily large cost, N. Let $\overline{m} = -M$. Then the maximum $\overline{M}$ of the artificial problem is

$$\overline{M} = M - N(x_{-10} + x_{-11} + x_{-12} + x_{-13} + x_{-14})$$
$$\overline{M} + \overline{m} + N(x_{-10} + x_{-11} + x_{-12} + x_{-13} + x_{-14}) = 0.$$

Using our simplified notation, the initial condensed tableau for the Simplex Method is

	1	2	3	4	5	6	7	8	9	
−10	1	1	1	0	0	0	0	0	0	10
−11	0	0	0	1	1	1	0	0	0	25
−12	0	0	0	0	0	0	1	1	1	20
−13	1	0	0	1	0	0	1	0	0	15
−14	0	1	0	0	1	0	0	1	0	18
15	2	1	5	7	4	3	6	2	4	0
16	−2	−2	−1	−2	−2	−1	−2	−2	−1	−88

The objective function has been given a double row with the coefficients of N in the bottom row. Each of the first five rows has been subtracted from the bottom row in order to get the artificial variables into the basis. The tableau is ready to run. A run on the computer produces the same answer that we arrived at in section 10.7. However, the computer takes 10 pivots compared to the two iterations of the Transportation Algorithm. The Simplex tableau is 7×10 compared to 3×3. The difference becomes more striking as the size of the problem increases. For example in a transportation problem with a 10×10 cost matrix the Simplex tableau mushrooms to 21×101. For a sizable problem it is not difficult to exceed the capacity of your machine. ●

Example 5. As a final technique let us solve the same problem in Example 4 without the use of artificial variables. The artificial variables may be avoided by finding an initial basic feasible solution for the problem variables. Examine the initial tableau above and note that columns 3, 6, and 9 are already unit vectors except for the objective rows. To get two more columns in the same condition subtract the 5th row from the 2nd row and the 4th row from the 3rd row. The result is the following tableau where columns 3, 5, 6, 7, and 9 are unit vectors except for the objective row.

1	2	3	4	5	6	7	8	9	
1	1	1	0	0	0	0	0	0	10
0	−1	0	1	0	1	0	−1	0	7
−1	0	0	−1	0	0	0	1	1	5
1	0	0	1	0	0	1	0	0	15
0	1	0	0	1	0	0	1	0	18
2	1	5	7	4	3	6	2	4	0

A basic solution may be completed by subtracting from the objective row 5 times the first row, 3 times the second row, 4 times the third row, 6 times the fourth row, and 4 times the fifth row. This step gives the following extended tableau with a basic feasible solution.

1	2	3	4	5	6	7	8	9	
1	1	1	0	0	0	0	0	0	10
0	−1	0	1	0	1	0	−1	0	7
−1	0	0	−1	0	0	0	1	1	5
1	0	0	1	0	0	1	0	0	15
0	1	0	0	1	0	0	1	0	18
−5	−5	0	2	0	0	0	−3	0	−253

The basic feasible solution read off from the unit vectors is $x_3 = 10$, $x_6 = 7$, $x_9 = 5$, $x_7 = 15$, $x_5 = 18$, and $\bar{m} = -M = 253$. The extended tableau may now be condensed by dropping the columns of unit vectors to get an initial tableau for a machine run with our automatic routine.

	1	2	4	8	
3	1	1	0	0	10
6	0	−1	1	−1	7
9	−1	0	−1	1	5
7	1	0	1	0	15
5	0	1	0	1	18
	−5	−5	2	−3	−253

A run of this condensed tableau produces the same familiar solution of cost 155, but the machine requires only 3 pivots. In addition to greatly reducing the number of pivots, the initial tableau size has been cut down to 6 × 5. The reduction accomplished here might be enough in a large problem to get the tableau down to machine size. ●

Example 6. A very interesting comparison can be made if, instead of the initial tableau in Example 5, we start with the initial condensed tableau corresponding to the Northwest Corner solution. In order to find this tableau let us start again with the initial simplex tableau found from the row and column requirements

1	2	3	4	5	6	7	8	9	
1	1	1	0	0	0	0	0	0	10
0	0	0	1	1	1	0	0	0	25
0	0	0	0	0	0	1	1	1	20
1	0	0	1	0	0	1	0	0	15
0	1	0	0	1	0	0	1	0	18
2	1	5	7	4	3	6	2	4	0

Recall that variables x_1, x_4, x_5, x_6, and x_9 are in the Northwest Corner solution. We must produce unit vectors in these five columns. Subtract the first row from the fourth row. Next add the fourth and fifth rows together, and then subtract their sum from the second row. The result shown in the following tableau has unit vectors in columns 1, 4, 5, 6, and 9 with the exception of the objective row.

1	2	3	4	5	6	7	8	9	
1	1	1	0	0	0	0	0	0	10
0	0	1	0	0	1	−1	−1	0	2
0	0	0	0	0	0	1	1	1	20
0	−1	−1	1	0	0	1	0	0	5
0	1	0	0	1	0	0	1	0	18
2	1	5	7	4	3	6	2	4	0

The Northwest Corner basic feasible solution may be completed by subtracting from the objective row 2 times the first row, 7 times the fourth row, 4 times the fifth row, 3 times the second row, and 4 times the third row. The result of these elementary row operations is the initial extended tableau:

1	2	3	4	5	6	7	8	9	
1	1	1	0	0	0	0	0	0	10
0	0	1	0	0	1	−1	−1	0	2
0	0	0	0	0	0	1	1	1	20
0	−1	−1	1	0	0	1	0	0	5
0	1	0	0	1	0	0	1	0	18
0	2	7	0	0	0	−2	−3	0	−213

Dropping the columns of unit vectors and placing the corresponding variables in the basis, gives the initial condensed tableau:

	2	3	7	8	
1	1	1	0	0	10
6	0	1	−1	−1	2
9	0	0	1	1	20
4	−1	−1	1	0	5
5	1	0	0	1	18
	2	7	−2	−3	−213

Compare this initial condensed tableau with the initial solution matrix found by the Transportation Algorithm in section 10.7. Especially note that the shadow prices above agree with the entry costs in the Transportation Algorithm. The solution variables have the same values and the cost of this solution, $m = -M = 213$, is the same in both cases.

We continue the Simplex Method by pivoting the condensed tableau. The first pivot is at position $(P, Q) = (5, 4)$, and the result is found in the next tableau.

	2	3	7	5	
1	1	1	0	0	10
6	1	1	−1	1	20
9	−1	0	1	−1	2
4	−1	−1	1	0	5
8	1	0	0	1	18
	5	7	−2	3	−159

Compare this tableau with the second matrix in the Transportation Algorithm solution. Again everything agrees including the shadow prices with the entry costs. The final pivot is at $(P, Q) = (3, 3)$, and we now look at the final tableau.

	2	3	9	5	
1	1	1	0	0	10
6	0	1	1	0	22
7	−1	0	1	−1	2
4	0	−1	−1	1	3
8	1	0	0	1	18
	3	7	2	1	−155

The shadow prices or entry costs are all positive so the cost of the minimum solution is 155. Our comparison is complete. The sequence of solutions found by the Simplex Method is identical to the sequence of solutions found by the Transportation Algorithm. The two methods are equivalent provided only that we start from the same initial basic feasible solution. ●

The simplifying feature of the Transportation Problem is the fact that the coefficients of all of its constraints are zeros or ones. The Transportation Algorithm takes advantage of this feature and thus makes a marked improvement over the general Simplex Method. However our final method in Example 6 above made similar use of the zeros and ones by finding basic feasible solutions from elementary row operations. Such solutions can always be found when the tableau coefficients are mostly zeros with the rest ones. If we are clever enough to find the same initial solution as found by the Northwest Corner Method, then our final method shares all of the advantages of the Transportation Algorithm and gives us the optimal solution in precisely the same number of identical steps.

We may make use of this analogy in many ways. For example, since entry costs are the same as shadow prices, parametric programming can be carried out on the Transportation Problem in much the same way as in Chapter 9. As a final thought in our presentation of linear programming let us reemphasize the beauty and simplicity of the Simplex Method both as a theoretical model and as a practical tool in solving a vast variety of linear problems.

Problems Chapter 10

Solve the following transportation problems by the Transportation Algorithm. Perturb if necessary. Indicate those problems that are degenerate and those that have alternate optima. Give an additional optimal solution in cases of alternate optima. In each problem the cost matrix is given along with the row and column requirements.

10–1.

		destinations 1	2	3	
origins	1	45	32	50	100
	2	74	65	81	150 row requirements
	3	28	47	56	75
		90	120	115	

column requirements

10–2. The following is the first published transportation problem.[2]

		destinations 1	2	3	4	
	1	10	5	6	7	25
origins	2	8	2	7	6	25
	3	9	3	4	8	50
		15	20	30	35	

10–3.

		destinations 1	2	3	
	1	7	8	9	150
origins	2	5	10	2	80
	3	4	6	3	170
	4	8	5	12	50
		100	200	150	

10–4.

		destinations 1	2	3	4	
	1	2	3	1	4	30
origins	2	5	7	8	9	80
	3	6	9	2	5	50
	4	4	10	6	8	70
		75	60	45	50	

10–5. Solve Problem 10–4 if variables X_{24} and X_{42} are required to be in the solution by changing their cost coefficients to zero. Does this change increase both of these variables to their maximum possible value?

[2] Frank L. Hitchcock, *Journal of Mathematics and Physics*, 20 (1941): 224–230.

10–6.

		destinations					
		1	2	3	4	5	
origins	1	5	3	8	7	5	110
	2	7	6	4	5	3	90
	3	8	9	2	4	6	120
		55	40	90	75	60	

10–7. Solve Problem 10–6 and determine the minimum cost if the third origin is not permitted to ship to the third destination.

10–8. In Problem 10–6 suppose the entries of the cost matrix represent ratings instead of costs. Solve the problem so as to achieve the maximum possible rating.

10–9.

		destinations				
		1	2	3	4	
origins	1	10	22	16	24	1000
	2	12	14	28	36	1500
	3	15	20	35	18	1500
		950	890	900	925	

Note an additional column must be added since supply exceeds demand. How many items will not be shipped from each of the three origins?

10–10. Find the maximum solution to Problem 10–9.

10–11.

		destinations				
		1	2	3	4	
origins	1	10	6	4	8	1000
	2	8	12	3	10	1000
	3	7	5	9	2	1000
		800	960	780	850	

In this case the demand is greater than the supply so that an additional row must be added. For the optimum solution, what are the shortages at each of the four destinations?

10–12. Find the maximum solution to Problem 10–11. Also note the shortages in this case.

The remaining problems in this chapter should be solved by the Simplex Method using the automatic routine developed in Chapter 4. In each case give the solution matrix and the optimum value of the objective function.

10–13. Solve the following transportation problem for minimum cost.

		destinations						
		1	2	3	4	5	6	
origins	1	2.3	3.2	1.6	5.4	4.8	2.7	5000
	2	3.1	5.2	3.6	7.1	2.5	1.9	4000
	3	4.4	6.0	1.5	6.3	3.3	1.7	6000
		2500	2250	2320	980	1350	5600	

10–14. Solve Problem 10–13 for the maximum cost.

10–15. Solve the following transportation problem for minimum cost. Is the solution unique?

		destinations					
		1	2	3	4	5	
origins	1	2.4	4.1	3.3	1.6	5.2	70
	2	6.1	3.5	2.9	4.4	1.9	85
	3	2.8	5.3	4.3	3.6	2.6	90
	4	3.2	6.2	1.7	2.8	4.1	65
		42	86	55	67	60	

10–16. Solve Problem 10–15 for the maximum cost.

10–17. Solve Problem 10–13 without the use of artificial variables by first finding a basic feasible solution as shown in section 10.10.

10–18. Solve Problem 10–15 without the use of artificial variables.

10–19. The Flash delivery service has trucks at three locations around town. This morning orders for pick up came in from six different customers. The dispatcher compiles the following mileage chart to show the distance from each truck garage to the various customers:

		customer							
		1	2	3	4	5	6		
garages	1	10	8.5	12	9.0	3.0	1.5	10	delivery trucks
	2.	3.5	4.0	6.0	5.5	7.0	3.5	8	
	3	15	6.5	11	13	7.5	4.5	11	
	trucks needed	1	2	4	1	3	5		

How should the dispatcher detail his delivery trucks to minimize his cost assuming that the cost is proportional to mileage? What is the minimum required mileage?

10–20. An accident completely ties up one road so that the trucks in Problem 10–19 from garage two cannot get through to the third customer. What is the new solution assuming that the dispatcher can reroute his trucks?

10–21. The Seeall Company has two principal factories that produce its color TV sets. The sets are sold through seven distributors around the country. Factory *A* has a capacity of 5000 sets per month and the capacity of factory *B* is 7000 sets per month. Next month the requirements of the seven distributors are respectively 1000, 1250, 1850, 900, 2200, 1500, and 2450 TV sets. The TV sets are shipped in lots of 10. The costs per lot of shipping to the distributors are summarized in the following table:

		distributors						
		1	2	3	4	5	6	7
factories	*A*	$54	62	85	40	105	68	88
	B	$73	58	75	90	73	67	92

What is the required production at each factory next month and what is the shipping schedule to each distributor that will minimize the month's shipping costs?

10–22. A hurricane sweeps inland from the coast leaving a path of destruction with telephone and power lines down in six communities. In four nearby cities the telephone and electric power companies set up emergency crews of linemen to send to the stricken areas. It is essential to get power and communications restored as soon as possible. The linemen travel in crews of two men each. Some crews will be airlifted in by helicopter while others will have to travel by truck. City *A* has 14 men ready to go in 7 helicopters with pilots. Cities B, C, and D have respectively 18, 14, and 8 men ready to go in their service trucks. The six stricken communities are in need of 6, 10, 8, 12, 6, 12 linemen respectively. The following time chart gives the number of minutes necessary for each crew to reach one of the possible destinations by its mode of travel.

		stricken communities					
		1	2	3	4	5	6
cities	A	18	35	60	24	36	42
	B	40	65	38	70	34	80
	C	45	38	48	55	50	76
	D	20	72	25	60	54	94

How should the telephone and electric power companies allocate line crews so as to minimize the total crewminutes of time to reach the disaster areas?

10–23. (*Optimal Assignment Problem*) An executive has four department head positions to be filled. His personnel manager has screened out four men to be hired. Each of the men is given a rating on a scale from 0 to 100 for each of the four jobs. A rating of 100 means that the man is perfectly suited for the job. The ratings are given in the following table:

		job			
		A	B	C	D
man	1	70	80	75	90
	2	75	60	85	78
	3	82	76	84	88
	4	80	68	77	82

Determine which man goes to which job so that the executive achieves the highest total rating for his four departments.

10–24. A machine shop has an order for a number of articles, each of which is composed of four parts that are to be made on four different machines. Four machinists are the employees that run the various machines. From previous work the shop foreman has composed the following table of costs for each part made by each of the machinists. The costs are given in dollars and cents.

		parts			
		A	B	C	D
machinists	1	2.40	4.60	3.55	7.90
	2	2.50	4.35	3.25	8.30
	3	2.50	4.25	3.60	8.15
	4	2.45	4.10	3.50	7.75

How should the foreman assign his machinists to the jobs so as to minimize the cost of the final product? What is that optimal cost?

10–25. (*Warehouse Problem*[3]) The warehouse problem was first published by Albert S. Cahn in 1948. It is not a transportation problem but rather a general linear programming problem. It is similar to the transportation problem in that the initial tableau is composed of coefficients that are all zeros and ones or minus ones. This means that all tableaux throughout the iterations will be integral. Solve the following example of a warehouse problem.

A certain warehouse will hold 1000 tons of grain. At the beginning of the month there are 600 tons of grain in the warehouse. The manager wishes to buy and sell grain on a weekly basis for the month. His estimated costs C_j per ton and selling prices P_j per ton for the 4 weekly periods $j = 1,2,3,4$, are given in the following table.

	weeks			
	1	2	3	4
C_j	\$150	155	125	130
P_j	\$170	165	150	140

It is assumed that markets are available so that sales and purchases can be of the desired size at the prevailing P_j and C_j. Sales must be made from the inventory on hand at the beginning of each weekly period. Of course the amount purchased during any week is limited by the capacity of the warehouse. The problem is to determine the optimal plan of purchasing, storage, and sales so as to maximize the gross profit. Hint: Let X_j, $j = 1, \ldots, 4$ be the number of tons to be purchased in week j. Let X_j, $j = 5, \ldots, 8$

[3] Albert S. Cahn, "The Warehouse Problem," *Bulletin of the American Mathematical Society*, 54 (November 1948): 1073.

be the number of tons to be sold in week $(j - 4)$. Then the buying constraints due to the capacity of the warehouse are:

$$\sum_{j=1}^{i} X_j - \sum_{j=5}^{i+4} X_j \leq 1000 - 600 = 400, \text{ for each } i = 1, \ldots, 4.$$

The selling constraints due to the amount on hand at the beginning of each week are:

$$-\sum_{j=1}^{i-1} X_j + \sum_{j=5}^{i+4} X_j \leq 600, \text{ for each } i = 1, \ldots, 4.$$

The nonnegativity constraints are:

$$X_j \geq 0, j = 1, \ldots, 8.$$

The linear function to be maximized is:

$$M = \sum_{j=1}^{4} -C_j X_j + \sum_{j=5}^{8} P_j X_j.$$

In general if there are n periods the matrix of constraint coefficients is $2n \times 2n$ and is made up of zeros and plus or minus ones.

10–26. Solve Problem 10–25 if the costs and selling price per ton are given in the following table.

	weeks 1	2	3	4
C_j	\$150	140	135	142
P_j	\$148	145	150	140

REFERENCES

The following references are restricted to books that are either on linear programming or related subjects covered in this text. Long lists of journal articles and research papers on specific topics may be found in the reference lists of S. I. Gass or T. C. Hu.

1. W. J. Adams, A. Gewirtz and L. V. Quintas, "Elements of Linear Programming," (New York: Van Nostrand Reinhold, 1969).
2. E. M. L. Beale, "Mathematical Programming," (London: Isaac Pitman and Son, 1968).
3. A. Charnes and W. W. Cooper, "Management Models and Industrial Applications of Linear Programming," (New York: John Wiley and Sons, 1961).
4. An-min Chung, "Linear Programming," (Columbus, Ohio: Charles E. Merrill, 1963).
5. L. Cooper and D. Steinberg, "Introduction to Methods of Optimization," (Philadelphia: W. B. Saunders, 1970).
6. G. B. Dantzig, "Linear Programming and Extensions," (Princeton, N.J.: Princeton University Press, 1963).
7. N. J. Driebeek, "Applied Linear Programming," (Reading, Mass.: Addison-Wesley, 1969).
8. R. J. Duffin, E. L. Peterson and C. Zener, "Geometric Programming—Theory and Application," (New York: John Wiley and Sons, 1967).
9. J. R. Frazer, "Applied Linear Programming," (Englewood Cliffs, N.J.: Prentice-Hall, 1968).

10. D. Gale, "The Theory of Linear Economic Models," (New York: McGraw-Hill, 1960).

11. W. W. Garwin, "Introduction to Linear Programming," (New York: McGraw-Hill, 1960).

12. Saul I. Gass, "Linear Programming," third edition, (New York: McGraw-Hill, 1969).

13. A. M. Glicksman, "An Introduction to Linear Programming and the Theory of Games," (New York: John Wiley and Sons, 1963).

14. R. L. Graves and P. Wolfe, "Recent Advances in Mathematical Programming," (New York: McGraw-Hill, 1963).

15. G. Hadley, "Linear Programming," (Reading, Mass.: Addison-Wesley, 1962).

16. T. C. Hu, "Integer Programming and Network Flows," (Reading, Mass.: Addison-Wesley, 1969).

17. S. Karlin, "Mathematical Methods and Theory of Games, Programming, and Economics," vols. I and II, (Reading, Mass.: Addison-Wesley, 1959).

18. L. S. Lasdon, "Optimization Theory of Large Systems," (New York: Macmillan, 1970).

19. R. W. Llewellyn, "Linear Programming," (New York: Holt, Rinehart, Winston, 1964).

20. J. C. C. McKinsey, "Introduction to the Theory of Games," (New York: McGraw-Hill, 1952).

21. Kurt Meisels, "A Primer of Linear Programming," (New York: New York University Press, 1962).

22. John von Neumann and O. Morgenstern, "Theory of Games and Economic Behavior," science editions, (New York: John Wiley and Sons, 1964).

23. Guillermo Owen, "Game Theory," (Philadelphia: W. B. Saunders, 1968).

24. Guillermo Owen, "Finite Mathematics," (Philadelphia: W. B. Saunders, 1970).

25. M. Simonnard, "Linear Programming," (Englewood Cliffs, N.J.: Prentice-Hall, 1966).

26. W. Allen Spivey, "Linear Programming, An Introduction," (New York: Macmillan, 1966).

27. S. Vajda, "Mathematical Programming," (Reading, Mass.: Addison-Wesley, 1961).

28. F. A. Valentine, "Convex Sets," (New York: McGraw-Hill, 1964).

29. D. J. Wilde and C. S. Beightler, "Foundations of Optimization," (Englewood Cliffs, N.J.: Prentice-Hall, 1967).

30. S. I. Zukhovitskiy and L. I. Avdeyeva, "Linear and Convex Programming," (Philadelphia: W. B. Saunders, 1966).

APPENDIX A

FORTRAN Subroutines for Primal-Dual Algorithm

In Chapter 4 a set of subroutines was given for the automatic pivoting of a linear programming tableau where artificial variables, if present, were removed from the basis. This appendix contains an alternate set of subroutines for the automatic pivoting of an initial tableau. The technique programmed here avoids artificial variables by the methods of Chapter 7. The following subroutines are to be used with a program that follows the Primal-Dual Algorithm for Mixed Systems given in Section 7.2.

The initial tableau is to be set up according to the first four steps of the algorithm in Section 7.2. Thus, all of the constraints have been converted to type I inequalities ($\leq$), and the objective function is a function to be maximized. The resulting initial tableau may be infeasible or nonoptimal or both. All possible pivots will be considered by both the Simplex Method and the Dual Simplex Method.

To carry out the algorithm's fifth step, first call SUBROUTINE DROW.

```
      SUBROUTINE DROW (Y,M,N,IP,W)
      DIMENSION Y(30,30),W(30)
      IP=-1
      B=0.
      MM=M-1
      DO 935 I=1,MM
      IF(Y(I,N)+.1E-5) 915,935,935
  915 IF(W(I)) 920,920,935
  920 IF(Y(I,N)-B) 925,935,935
  925 B=Y(I,N)
      IP=I
  935 CONTINUE
      RETURN
      END
```

The vector W is to be initialized by the main program at W(I) = 0 for I = 1 to M. It will be used to skip the Ith row if that row has the most negative number in its last column but no possible pivot. Then the next most negative number in the last column will determine a possible pivotal row. The first IF statement in DROW is for the purpose of skipping a row with either zero or a positive number in its last column. Note the epsilon that has been added as a bow to the inaccuracies of the computer. If Y(I,N) should be zero but turns up as a small negative number due to round-off error, then the small positive number will correct the problem of choosing that row. Statement 915 causes the Ith row to be skipped whenever W(I) is positive. Statements 920 and 925 are similar to the corresponding statements in SUBROUTINE COL and cause the most negative number available, Y(I,N), to pick the pivotal row, IP. Of course, if there is no possible pivotal row, IP is returned to the main program at −1. If DROW returns a positive IP, then that value should be stored for future reference and SUBROUTINE DCOL called.

```
         SUBROUTINE DCOL (Y,M,N,IQ,IP,DS)
         DIMENSION Y(30,30)
         IQ=-1
         DS=1.E19
         NN=N-1
         DO780 J=1,NN
         IF(Y(IP,J)+.1E-5) 740,780,780
     740 IF(Y(M,J)+.1E-5) 780,750,750
     750 R=-Y(M,J)/Y(IP,J)
         IF(DS-R) 780,780,760
     760 DS=R
         IQ=J
     780 CONTINUE
         RETURN
         END
```

This subroutine looks for a pivotal column by examining the $\bar{\theta}$ ratios between the last row and the pivotal row. The first IF statement will skip any zero or positive entry in the pivotal row. Again note the small positive number that has been added to take care of round-off error. For a negative entry in the pivotal row, statement 740 will skip that column if there is a negative number in the bottom row. Thus, the $\bar{\theta}$ ratios will be negative or zero. Statement 750 computes the positive value of the $\bar{\theta}$ ratios. The next two statements are similar to the corresponding statements in SUBROUTINE ROW. They pick the $\bar{\theta}$ ratio of smallest numerical value. Then IQ is set equal to the number of that column. If there is no possible pivotal column, then IQ = −1 is returned. In this case the main program should set W(IP) = 1 and call DROW again. Thus row IP will be skipped as we try to find another pivotal row.

This search is continued until both IP and IQ return positive or else there is no dual pivot. If no pivot has been found, proceed directly to step six of our algorithm. When a pivot has been determined, store

IP and IQ in the memory. Then compute the decrease in objective value by the formula

$$D_q = ABS(Y(IP, N) * DS).$$

The DS is the smallest numerical value of the $\bar{\theta}$ ratios found by DCOL in determining the pivotal column. We are now ready for step six of the Primal-Dual Algorithm.

In order to find a pivot by the Simplex Method, we use two subroutines that are only slightly different from COL and ROW found in Chapter 4. The first is COL2 that will search for a pivotal column.

```
      SUBROUTINE COL2 (Y,M,N,IQ,V)
      DIMENSION Y(30,30),V(30)
      IQ=-1
      B=0.
      NN=N-1
      DO 30 J=1,NN
      IF (V(J)) 10,10,30
   10 IF(Y(M,J)+.1E-5) 20,30,30
   20 IF (Y(M,J)-B)25,30,30
   25 B=Y(M,J)
      IQ=J
   30 CONTINUE
      RETURN
      END
```

The vector V(J) should initially be set equal to zero for J = 1 to N. It is used to skip column J when no pivot is available in column J. Its roll is analogous to that of W(I) in DROW. Column J will be skipped whenever V(J) is positive. In statement 10 an epsilon has been added to take care of round-off error. The same correction could be added to COL in Chapter 4. The remainder of COL2 is identical to COL.

If IQ = −1 is returned by COL2, then the pivot chosen by step five above will be used or else the pivoting is completed. If IQ returns positive it will be stored and subroutine ROW2 called.

```
      SUBROUTINE ROW2 (Y,M,N,IP,IQ,S)
      DIMENSION Y(30,30)
      IP=-1
      S=1.E19
      MM=M-1
      DO 80 I=1,MM
      IF(Y(I,IQ)-.1E-5) 80,80,5
    5 IF(Y(I,N)+.1E-5) 80,10,10
   10 R=Y(I,N)/Y(I,IQ)
      IF(S-R)80,80,20
   20 S=R
      IP=I
   80 CONTINUE
      RETURN
      END
```

The first IF statement will skip any negative or zero entries in the pivotal column. Notice that the epsilon value has been subtracted. This will eliminate any rows with zero values that turn out to be positive values due to round-off error. The same correction could be made to ROW in Chapter 4. For the positive entries in the pivotal column, statement 5 will skip those rows with a negative number in the last column. Again the epsilon correction has been added so that we do not skip any negative numbers that should be zero. Thus, the θ ratios considered will be positive or zero. The remaining statements that pick the minimum θ ratio and set IP are identical to those of ROW.

If ROW2 returns IP negative then set V(IQ) = 1 and call COL2 again. Since column IQ will be skipped, the next possible pivotal column will be considered before returning to ROW2. Continue this procedure until either no pivot is found or else both IP and IQ return positive. In case IP and IQ are both positive, compute the increase in objective value by the formula

$$I_p = ABS(Y(M,IQ)*S).$$

S is the minimum value of the considered θ ratios. If no pivot were found by the Dual Simplex Method, D_q could have been set equal to -1. Now compare I_p with D_q. Pick the larger of the two and set the pivot position (IP,IQ) equal to the values that were stored by the corresponding method. A call to SUBROUTINE PIVCO reproduced below will carry out a pivoting iteration at the chosen position.

```
      SUBROUTINE PIVCO (Y,K,L,M,N,IP,IQ)
      DIMENSION Y(30,30), K(30), L(30)
      R=1./Y(IP,IQ)
      DO 10 J=1,N
   10 Y(IP,J)=R*Y(IP,J)
      Y(IP,IQ)=R
      DO 50 I=1,M
      IF (I-IP) 20,50,20
   20 DO 40 J=1,N
      IF (J-IQ) 30,40,30
   30 Y(I,J)=Y(I,J)-Y(I,IQ)*Y(IP,J)
   40 CONTINUE
      Y(I,IQ)=-R*Y(I,IQ)
   50 CONTINUE
      X=K(IQ)
      K(IQ)=L(IP)
      L(IP)=X
      RETURN
      END
```

After completing PIVCO the vectors W(I) and V(J) must be reset to zero before searching for the next pivot position. As soon as no new pivot can be found, the final tableau has been reached and should be printed out for interpretation. The analysis of the final tableau is found in Section 7.2 following the Primal-Dual Algorithm.

APPENDIX B

FORTRAN Subroutines for Gomory's Algorithm

This appendix contains additional subroutines that will carry out integer programming by Gomory's Algorithm. Gomory's Algorithm, found in Section 8.3, is a method for reducing an optimal feasible tableau with nonintegral basic variables to one with integral basic variables. We start with the same initial tableau that was set up in Appendix A for the Primal-Dual Algorithm. Then to produce an optimal feasible tableau, we call SUBROUTINE MULPLX.

SUBROUTINE MULPLX (Y,M,N,K,L) will not be reproduced here because it is essentially the entire program for the Primal-Dual Algorithm. By making the entire program of Appendix A into a subroutine, we may call for the optimal feasible tableau and then continue directly with Gomory's Algorithm. It is suggested that the optimal feasible tableau be printed to see if the basic variables are already integral. If not then a call to SUBROUTINE CUT will carry out the first two steps of Gomory's Algorithm.

```
        SUBROUTINE CUT (Y,M,N,II,Z)
        DIMENSION Y(30,30), Z(30)
        B=0.
        MM=M-1
        DO 200 I=1,MM
        R=Y(I,N)-IFIX(Y(I,N)+.01)
        IF(R-B) 200,200,100
    100 II=I
        B=R
    200 CONTINUE
        DO 250 J=1,N
        IF(Y(II,J)) 230,210,210
    210 Z(J)=IFIX(Y(II,J)+.01)-Y(II,J)
        GO TO 250
    230 Z(J)=IFIX(Y(II,J)+.01)-1.-Y(II,J)
    250 CONTINUE
        RETURN
        END
```

The first DO loop in SUBROUTINE CUT picks out the row with the largest fractional part in its right-hand column. This row number is set equal to II. For a positive number the function IFIX returns the next integer less than or equal to the given number. An epsilon of .01 has been added so that round-off error could not produce a lower integer than desired from IFIX.

The second DO loop finds the coefficients in the new row to be added to the optimal feasible tableau. These coefficients are called Z(J). There are two branches because IFIX operates differently on negative numbers than it does on positive numbers. In the BASIC language the function INT may be used for both cases. For a negative number IFIX returns the next integer that is larger than or equal to the given number. This will be the integer that is closer to the origin. Thus, in statement 230 one is subtracted from integer IFIX to move to the next integer away from the origin. In both cases note the order of subtraction so that Z(J) is negative or zero.

Step three of our algorithm is to increase the size of the current tableau by adding in the new row of coefficients just above the objective row. This is accomplished by calling SUBROUTINE ADD.

```
          SUBROUTINE ADD (Y,M,N,L,Z,MU)
          DIMENSION Y(30,30),L(30),Z(30)
          M=M+1
          M1=M-1
          L(M)=L(M1)
          L(M1)=MU+1
    ┌──   DO 300 J=1,N
    │     Y(M,J)=Y(M1,J)
    └─ 300 Y(M1,J)=Z(J)
          MU=MU+1
          RETURN
          END
```

The variable MU is set equal to L(M) by the main program before calling ADD. MU is increased by one each time ADD is called. The number of rows M is also increased by one each time. L(M1) is the subscript of the new slack variable assigned to the new row. Thus the new slack variable has the next higher integer for its subscript each time around. The DO loop moves the coefficients of the objective row into the Mth row and places the Z(J) into the M−1 row.

The tableau is now ready to pivot and a call to SUBROUTINE MULPLX gives the next optimal feasible tableau. Again the result should be printed to see if the basic variables are integral. If not go back to SUBROUTINE CUT and repeat the process. It may happen that one of the resulting tableaux is infeasible with no place to pivot. This would mean that the problem has no integral solution. The feasible region may not contain any points with integral coordinates. In that case all solutions will involve some basic variable that is nonintegral.

The process of adding cuts may be made automatic. This would be desirable in a problem requiring many cuts. After each cycle of the current program, test the basic variables to see if they are within some epsilon of an integer. If the test is not met, call for another cut. When the test is met, print out the final result. However, if some tableau is infeasible after a call to MULPLX, the process must be halted because there is no integer solution.

FORTRAN GLOSSARY

ABS SIGNIFIES THE ABSOLUTE VALUE FUNCTION. IT TAKES A REAL ARGUMENT AND DELIVERS A REAL ANSWER. SEE IABS.

ACCURACY. SEE PRECISION.

ALOG SIGNIFIES THE NATURAL LOGARITHM FUNCTION, I.E., THE LOGARITHM TO THE BASE 2.71828....

ARCTANGENT. SEE ATAN.

ARITHMETIC STATEMENT FUNCTION. FOR EXAMPLE

```
CUBRT(H)=H**(1./3.)
```

MAY BE PLACED IN A SOURCE DECK TO DEFINE THE CUBE-ROOT FUNCTION. THEN LATER IN THE PROGRAM ONE MAY FORM EXPRESSIONS LIKE

```
1.2+CUBRT(2.-FOD)
```

HERE THE VARIABLE H IS A DUMMY, STANDING FOR EXPRESSIONS TO BE PLUGGED IN LATER. ONE MAY USE MORE THAN ONE ARGUMENT. FOR EXAMPLE

```
PYTHA(X,Y)=SQRT(X*X+Y*Y)
```

OR EVEN

```
JSUM(I,J,K,L,M)=I+2*J+3*K+4*L+5*M
```

THE MAXIMUM NUMBER OF DUMMY ARGUMENTS FOR AN ARITHMETIC STATEMENT FUNCTION IS FIFTEEN.
ALSO, IT IS LEGAL FOR AN ARITHMETIC STATEMENT FUNCTION TO USE ANY FUNCTION PREVIOUSLY DEFINED, IN OR OUT OF THE PROGRAM. FURTHERMORE, ONE MAY REFER TO VARIABLES OTHER THAN THE DUMMIES. FOR EXAMPLE

```
SQUAR(Y)=Y*Y
GAUSS(B)=EXP(.5*SQUAR(B))/SQRT(2.*PI)
PI=4.*ATAN(1.)
```

CARDS DEFINING ARITHMETIC STATEMENT FUNCTIONS COME AFTER CHANGE OF TYPE, DIMENSION, COMMON, AND EQUIVALENCE CARDS, BUT BEFORE FORMATS AND EXECUTABLE CARDS.

ARRAY. IF A VARIABLE HAS BEEN DIMENSIONED THEN WE CALL IT AN ARRAY, OR SOMETIMES A DIMENSIONED VARIABLE. FOR EXAMPLE THE CARD

```
DIMENSION Y(20,20)
```

DECLARES Y TO BE A SQUARE ARRAY 20 ON A SIDE. THE ARGUMENTS OF A CALL OR FUNCTION MAY BE ARRAY NAMES INSTEAD OF EXPRESSIONS, SO IN THE SAME PROGRAM ONE MIGHT SEE

```
CALL PIVOT(Y,I,J,5,3)
```

WHERE PIVOT IS A SUBROUTINE PLACED IN MEMORY AS PART OF THIS JOB. FOR THIS TO WORK PROPERLY, THE FIRST ARGUMENT OF THE PIVOT SUBROUTINE WILL HAVE TO BE A REAL SQUARE ARRAY 20 BY 20.

ASTERISKS HAVE THREE USES IN FORTRAN. ONE ASTERISK IS THE SIGN FOR MULTIPLICATION, TWO ASTERISKS TOGETHER MEAN TAKE A POWER, SO

```
WEIRD**TALES
```

MEANS RAISE WEIRD TO THE TALES POWER, AND LAST BUT NOT LEAST A FIELD OF ASTERISKS IN YOUR OUTPUT WHERE YOU EXPECTED A NUMBER MEANS THAT YOUR FORMAT WAS TOO SMALL TO HOLD THAT NUMBER. THE ONLY CURE FOR THIS IS TO USE A BIGGER FORMAT. JUST TO BE ON THE SAFE SIDE, USE A MUCH BIGGER FORMAT. IF ONLY TWO NUMBERS ARE TO COME ON A PRINTED LINE, TRY 2F60.8, BECAUSE HOW COULD YOU SAVE PAPER WITH A NARROWER FORMAT?

ATAN SIGNIFIES THE ARCTANGENT FUNCTION. THE ANSWER WILL BE IN RADIANS, BETWEEN MINUS PI OVER TWO AND PLUS PI OVER TWO.

BINARY. THE COMPUTER IS REALLY IN THE BASE 2, NOT 10, BUT IT IS NOT USUALLY NECESSARY TO THINK ABOUT THIS, BECAUSE FORTRAN MAKES ALL CHANGES BACK AND FORTH BETWEEN BASES 2 AND 10. HOWEVER, 10.*.1 IS NOT EQUAL TO 1.0, BUT IS A LITTLE LESS. THIS IS BECAUSE .1 CANNOT REALLY EXIST INSIDE A BINARY MACHINE.

BLANKS USUALLY MAY BE USED OR OMITTED TO SUIT YOUR OWN TASTE. COLUMN 6 IS BLANK EXCEPT FOR CONTINUATION

CARDS. BLANKS INSIDE A HOLLERITH FIELD, FOR INSTANCE 'HOWDY THERE FOLKS,' WILL BE PRINTED AS BLANKS.

BUFFER. ALL FORTRAN INPUT AND OUTPUT IS BY WAY OF A PART OF MEMORY CALLED A BUFFER. THE GREATEST ALLOWED BUFFER SIZE IS 80 FOR CARDS BOTH IN AND OUT, 80 FOR READING THE CONSOLE KEYBOARD, 120 FOR TYPING AT THE CONSOLE, AND 1 PLUS 120 FOR THE LINE PRINTER. THE EXTRA 1 IS THE CONTROL CHARACTER. AN ATTEMPT TO TAKE MORE THAN THE ALLOWED BUFFER WILL CAUSE THE PROGRAM TO FALL DEAD AND DISPLAY LIGHTS IN THE ACCUMULATOR.

CALL. FOR INSTANCE

```
CALL PIVOT(Y,I,J,6,8)
```

WHERE PIVOT IS THE NAME OF A SUBROUTINE AND PIVOT TAKES FIVE ARGUMENTS. THE CALL AND THE SUBROUTINE MUST AGREE IN NUMBER AND TYPE OF ARGUMENTS AND THEIR DIMENSIONALITY. THAT IS, IF Y IS DIMENSIONED A CERTAIN WAY IN THE PROGRAM THAT CALLS, THEN THE CORRESPONDING DUMMY IN THE SUBROUTINE MUST BE DIMENSIONED JUST THE SAME WAY. ARGUMENTS ON THE CALL CARD MAY BE EITHER EXPRESSIONS OR ARRAY NAMES.

CARRIAGE CONTROL. SEE SPACE.

CHANGE-OF-TYPE CARD. SEE TYPE.

CHARACTERS ARE LETTERS OR DIGITS OR OTHER MARKS OR THE BLANK. A CARD HAS 80 CHARACTERS.

COLUMN. A FORTRAN SOURCE STATEMENT USUALLY BEGINS IN COLUMN 7 AND DOES NOT PASS BEYOND COLUMN 72. COLUMN 6 SHOULD BE BLANK, EXCEPT FOR CONTINUATIONS. STATEMENT NUMBERS GO IN COLUMNS 1 THROUGH 5. COLUMN 1 IS FOR THE C OF A COMMENT CARD. THESE RULES HAVE NOTHING TO DO WITH DATA CARDS. THE LATTER CAN HAVE ALL 80 COLUMNS USED ACCORDING TO YOUR FORMAT SPECIFICATIONS.

COMMAS SHOWN IN THE EXAMPLES ARE MOSTLY MANDATORY. THE ONLY OPTIONAL COMMAS IN FORTRAN ARE THE COMMAS IN A FORMAT CARD FOLLOWING A HOLLERITH FIELD OR AN X FORMAT. EXAMPLES ARE

```
5 FORMAT ('□',F20.8,3X,I13)
```
[1]

[1] □ Represents a blank

AND ALTERNATIVELY

5 FORMAT('□'F20.8,3XI13)

IT IS BETTER TO PUT ALL THE COMMAS IN, FOR CLARITY.

COMMENT CARDS HAVE A C IN COLUMN 1. THEY ARE NOT READ BY THE COMPILER. EXAMPLE

C THIS NEXT LOOP IS TO SEARCH FOR THE NEGATIVE ROOT.

COMMON. THE COMMON IS A PART OF MEMORY NOT BELONGING TO A SINGLE PROGRAM BUT INSTEAD BELONGING TO ALL THE PROGRAMS OF A JOB THAT ARE LOADED TOGETHER. FOR EXAMPLE

COMMON X,Y(3,3),J(2)

DIMENSIONS Y AND J AS SHOWN AND THEN PLACES X, Y, AND J IN THE COMMON IN THE ORDER X, Y, J. THIS CARD SHOULD BE DUPLICATED WITH THE DUPLICATOR BUTTON OF THE KEY-PUNCH AND PLACED IN ALL THE PROGRAMS THAT ARE TO BE LOADED TOGETHER. THIS EXACTITUDE IS GOOD PROGRAMMING PRACTICE BECAUSE THE NAMES OF THE VARIABLES ARE NOT REMEMBERED, ONLY THEIR ADDRESSES. COMMON CARDS COME AFTER DIMENSION CARDS BUT BEFORE EQUIVALENCE AND FUNCTION DEFINITIONS.

COMPUTED GO TO. FOR EXAMPLE

GO TO (45,6,147),LULU

EXECUTED WHEN LULU IS 1 WILL CAUSE A TRANSFER TO 45, AND IF LULU IS 2 TO STATEMENT 6, AND IF LULU IS 3 TO 147. IF LULU IS LESS THAN 1 OR MORE THAN 3 THE PROGRAM WILL LEAVE THE TRACK AND NOBODY KNOWS WHAT WILL HAPPEN. HERE IS A LONGER EXAMPLE

GO TO(1,2,3,4,5,88,7,8,9),MOM

OBSERVE THE COMMA BEFORE THE VARIABLE. THE VARIABLE MUST BE JUST A SIMPLE INTEGER VARIABLE WITHOUT A SUBSCRIPT, NOT AN EXPRESSION. INSIDE THE PARENTHESES ONE MAY USE ONLY STATEMENT NUMBERS, WHICH ARE OF COURSE CONSTANTS.

CONSOLE. THE CONSOLE IS THE SCIENCE-FICTION-LOOKING DESK WITH BUTTONS AND LIGHTS.

CONSTANTS ARE THINGS LIKE 1, 23.5, AND 12.5E−2, AS OPPOSED TO VARIABLES SUCH AS X, Y, FPRIM, J, K123, AND JBAR. IF A PROGRAM DOES NOT GO OFF THE TRACK OR HAVE SUBSCRIPT ERRORS, THE CONSTANTS WILL HAVE THE SAME VALUE ALL THE WAY THROUGH, BUT THE VALUES OF THE VARIABLES WILL USUALLY CHANGE QUITE A BIT.

CONTINUATION CARDS ARE CARDS HAVING A NON-BLANK NON-ZERO CHARACTER IN COLUMN 6. SEE STATEMENTS.

CONTINUE STATEMENTS EXIST SOLELY FOR THE PURPOSE OF HAVING STATEMENT NUMBERS FOR REFERENCE OR TRANSFER. FOR EXAMPLE

```
55 CONTINUE
```

THIS IS CHIEFLY USEFUL FOR ENDING DO LOOPS. SAY,

```
      DO 55 J=1,81
      DO 44 K=1,81
      IF(J-K)33,44,33
   33 X(J,K)=0.
   44 CONTINUE
   55 CONTINUE
```

CONTINUE STATEMENTS ARE EXECUTABLE BUT THEY TAKE NO SPACE IN MEMORY AND NO TIME TO EXECUTE, SO WHEN IN DOUBT USE THEM.

COS SIGNIFIES THE TRIGONOMETRIC COSINE FUNCTION. THE ARGUMENT MUST BE IN RADIANS.

DATA CARDS COME AFTER THE END CARD. ALL 80 COLUMNS MAY BE USED, ACCORDING TO YOUR FORMAT SPECIFICATION.

DEBUG MEANS FIND THE MISTAKES IN THE PROGRAM. EVEN EXPERTS HAVE BUGS IN HALF THEIR PROGRAMS. THE MOST OBVIOUS WAY TO DEBUG A PROGRAM IS TO INSERT A LOT MORE WRITE STATEMENTS, JUST AS A TEMPORARY MEASURE.

DECIMAL. SEE BINARY. DECIMAL POINTS ARE FOUND ONLY ON REAL NUMBERS. F20.8 MEANS THE NUMBER WILL OCCUPY 20 COLUMNS INCLUDING BLANKS ON LEFT AND RIGHT, AND THERE WILL BE 8 DECIMALS TO THE RIGHT OF THE POINT. IF THIS IS AN INPUT RECORD, YOU MAY PUNCH THE DECIMAL POINT WHERE YOU PLEASE AND OVERRIDE THE FORMAT.

DIMENSION. IF X IS NOT A SINGLE VARIABLE BUT RATHER AN ARRAY OF NINE, THEN

```
      DIMENSION X(9)
```

MUST GO INTO THE SOURCE DECK AFTER THE CHANGE-OF-TYPE CARDS AND BEFORE THE COMMON AND EQUIVALENCE CARDS. LATER IN THE DECK ONE MIGHT WRITE STATEMENTS SUCH AS

```
      Y=2.*X(5)-23.3
```

OR PERHAPS

```
      DO 4 L=1,9
    4 X(L)=L+3
```

ONE CAN DEFINE MATRICES TOO.

```
DIMENSION X(9),Z(18),Q(9,18)
```

SHOWS HOW TO DEFINE TWO VECTORS (OF DIFFERENT LENGTHS) AND A MATRIX ON THE SAME CARD. THE MATRIX IS SAID TO HAVE 9 ROWS AND 18 COLUMNS. SEE THE ARTICLE ON SUBSCRIPTS. OF COURSE, INTEGER VARIABLES MAY BE DIMENSIONED ALSO.

DO STATEMENTS HAVE SPECIAL RULES. FIRST COMES THE WORD DO, THEN A STATEMENT NUMBER BELONGING TO A LATER CARD, THEN A NONSUBSCRIPTED INTEGER VARIABLE, THEN AN EQUAL SIGN, THEN TWO OR THREE INTEGER CONSTANTS OR NONSUBSCRIPTED INTEGER VARIABLES SEPARATED BY COMMAS. FOR EXAMPLE

```
DO 55 J=1,81
```

OR

```
DO 55 J=1,81,3
```

OR

```
DO 55 J=JGO,81
```

AND THE LIKE. THE STATEMENT NUMBER IS THE BOTTOM OF THE LOOP. THE VARIABLE FOLLOWING IS THE CONTROL VARIABLE OF THE LOOP. THEN WE SEE THE INITIAL VALUE OF THE CONTROL VARIABLE, AND ITS FINAL VALUE, AND THE STEP SIZE BY WHICH IT CHANGES. ALL THESE MUST BE STRICTLY POSITIVE, AND THE FINAL VALUE MUST NOT BE SMALLER THAN THE INITIAL, BUT THEY MAY BE EQUAL. IF THE STEP SIZE IS NOT GIVEN, THEN THE MACHINE ASSUMES YOU MEAN 1. HERE IS A SIMPLE DO LOOP.

```
    DO 55 J=1,81
    X(J)=0.
 55 Y(J)=0.
```

HERE IS A NEST OF THREE DO LOOPS.

```
    DO 55 I=1,2
    DO 55 J=1,2
    DO 55 K=1,2
 55 Z(I,J,K)=3.14159
```

THEY ARE ALLOWED, AS SHOWN HERE, TO END ON THE SAME CARD. IT IS FORBIDDEN TO END A DO LOOP ON AN IF OR GO TO OR COMPUTED GO TO OR A DO STATEMENT. IT IS ALSO FORBIDDEN TO END ON A FORMAT, BECAUSE A FORMAT IS NOT EVEN EXECUTABLE. THE CONTINUE STATEMENT (SEE CONTINUE) IS HANDY FOR ENDING DO LOOPS. IT IS LEGAL TO JUMP OUT OF A DO LOOP, BUT ILLEGAL TO JUMP IN FROM OUTSIDE. THIS ERROR IS NOT ALWAYS CAUGHT BY

THE MACHINE, BUT ONE SEES ON A MOMENT'S THOUGHT THAT THE INITIALIZATION MACHINERY WILL BE BYPASSED IN THIS CASE. IT IS ALSO ILLEGAL TO CHANGE THE VALUE OF THE CONTROL VARIABLE INSIDE THE LOOP. NESTED DO LOOPS MUST NOT HAVE THE SAME CONTROL VARIABLE.

DOUBLE-SPACE ON THE PRINTER IS CAUSED BY THE CONTROL CHARACTER ZERO. HENCE

```
      WRITE(6,14)
   14 FORMAT(1HO,'THESE ARE THE VALUES',///)
```

WILL CAUSE THE PRINTER TO TAKE A DOUBLE SPACE, PRINT 'THESE ARE THE VALUES' AND THEN TAKE 3 BLANK LINES.

E FORMAT IS RECOMMENDED FOR OUTPUT ONLY, BUT IT IS LEGAL FOR INPUT. FOR EXAMPLE THE FORMAT

```
   17 FORMAT(4E20.8)
```

DESCRIBES 4 REAL NUMBERS, EACH OCCUPYING 20 COLUMNS INCLUDING BLANKS, EACH HAVING 8 DECIMALS IN ITS MANTISSA TO THE RIGHT OF THE POINT, AND EACH HAVING A LETTER 'E' AND A SIGNED EXPONENT(BASE TEN). IF YOU HAVE NO IDEA AT ALL HOW BIG YOUR OUTPUT NUMBERS ARE, THIS IS A GOOD FORMAT TO USE, BECAUSE VERY LARGE NUMBERS WILL NOT TURN INTO FIELDS OF ASTERISKS.

END. EVERY PROGRAM SOURCE CARD DECK ENDS WITH THE CARD

```
      END
```

THIS CARD IS NOT EXECUTABLE, SO IT OUGHT NOT TO HAVE A STATEMENT NUMBER. THE END CARD IS NOT ABLE TO KILL THE PROGRAM WHEN IT IS FINISHED. FOR THIS ONE MAY USE THE CALL EXIT CARD OR THE STOP CARD.

EXIT. EXECUTION OF THE CARD

```
      CALL EXIT
```

WILL KILL THE PROGRAM WITHOUT EVEN BRINGING THE MACHINE TO A HALT. THIS IS GOOD ETIQUETTE FOR THE CLOSED SHOP.

EXP SIGNIFIES THE EXPONENTIAL FUNCTION. THIS GIVES 2.71828... TO THE ARGUMENT POWER. IT IS THE NATURAL ANTILOG.

EXPRESSION. FORTRAN EXPRESSIONS ARE MUCH LIKE THOSE OF ALGEBRA, BUT MULTIPLICATION MAY BE SHOWN ONLY BY THE USE OF THE ASTERISK, AND POWERS ARE SHOWN BY THE DOUBLE ASTERISK. FOR EXAMPLE

```
      AB+AC*(AD+AE)
```

MEANS ADD THE NUMBER AD TO AE AND MULTIPLY BY AC AND ADD TO AB. NOTICE THAT AB IS NOT A PRODUCT, BECAUSE THERE IS NO ASTERISK BETWEEN A AND B.

```
V**(1.2+GUG)
```

MEANS ADD 1.2 AND GUG TOGETHER AND THEN RAISE V TO THAT POWER. ACTUALLY THE FORMULA IS COMPILED INTO THE MACHINE THE SAME AS

```
EXP((1.2+GUG)*ALOG(V))
```

SO IF V IS NEGATIVE OR ZERO THE ANSWER WILL BE WRONG. OF COURSE FUNCTIONS AND DIMENSIONED VARIABLES MAY BE USED IN EXPRESSIONS, SO ONE SEES THINGS LIKE

```
X+1.2-4.5*FPRIM(X,Y)**U(J)
```

WHICH MIGHT MEAN CALCULATE FPRIM OF X AND Y, RAISE IT TO THE U SUB J POWER, MULTIPLY BY 4.5, AND SUBTRACT FROM THE SUM OF 1.2 AND X. WE SAY MIGHT MEAN, BECAUSE IT IS NOT CLEAR FROM THIS SNIPPET OF A PROGRAM WHETHER U(J) MEANS U SUB J OR THE VALUE OF THE U FUNCTION WHEN THE ARGUMENT IS J. THE PROGRAMMER IS WARNED THAT MULTIPLICATION AND DIVISION HAVE THE SAME PRECEDENCE RANK, SO THAT A FORMULA LIKE

```
A/B*C
```

WILL BE UNDERSTOOD BY THE MACHINE AS

```
(A/B)*C
```

AND NOT AS A/(B*C), WHICH IS PERHAPS WHAT WAS WANTED.

EXPRESSIONS CAN BE USED AS THE RIGHT-HAND SIDES OF REPLACEMENT STATEMENTS, AS THE RIGHT-HAND SIDES OF ARITHMETIC FUNCTION STATEMENTS, WITHIN IF STATEMENTS, AND AS ARGUMENTS OF FUNCTIONS AND CALLS TO SUBROUTINES. SUBSCRIPTS HAVE RESTRICTED RULES OF THEIR OWN, AND SO DO THE DO CARDS.

EXTERNAL FUNCTION. SEE FUNCTION.

F FORMAT IS THE FORMAT MOST OFTEN USED. FOR EXAMPLE

```
17 FORMAT(3F20.8)
```

DESCRIBES 3 REAL NUMBERS, EACH TAKING 20 COLUMNS INCLUDING BLANKS, AND EACH HAVING 8 DECIMALS TO THE RIGHT OF THE POINT. IF THIS FORMAT IS USED FOR OUTPUT THEN THE MACHINE WILL FOLLOW THE STYLE JUST DESCRIBED. ON INPUT, HOWEVER, ONE MAY BE A LITTLE DISOBEDIENT. ONE MAY, FOR EXAMPLE, HAVE SOME OTHER NUMBER OF DECIMALS TO THE RIGHT OF THE POINT, INSTEAD OF 8. OR ONE MAY OMIT THE POINT AND LET THE MACHINE PLACE IT BETWEEN THE FIRST 12 COLUMNS AND THE LAST 8

COLUMNS OF THE 20-COLUMN FIELD. SOME FORTRAN INPUT ROUTINES REQUIRE A NUMBER OF F TYPE TO BE RIGHT-JUSTIFIED IN ITS FIELD.

FIXED OR FIXED-POINT. SEE INTEGER.

FLOATING OR FLOATING-POINT. SEE REAL.

FORMATS CAN BE SIMPLE, FOR EXAMPLE

```
17 FORMAT(10F12.2)
```

WHICH MEANS 10 REAL-MODE NUMBERS EACH OCCUPYING 12 COLUMNS INCLUDING BLANKS, AND EACH HAVING 2 DIGITS TO THE RIGHT OF THE DECIMAL POINT. IF YOU ARE OUTPUTTING, THE MACHINE IS OBLIGED TO PREPARE THE NUMBERS JUST THAT STYLE. IF, ON THE CONTRARY, YOU ARE INPUTTING, YOU DO NOT HAVE TO FOLLOW FORMAT EXACTLY. AS LONG AS EACH NUMBER STAYS IN ITS ASSIGNED FIELD OF 12 COLUMNS, YOU MAY PUNCH THE DECIMAL POINT ANYPLACE IN THAT FIELD THAT YOU DESIRE. ALSO, YOU MAY OMIT THE POINT, BUT IF YOU DO, THE MACHINE WILL INSERT IT BETWEEN THE FIRST 10 COLUMNS OF THE FIELD AND THE LAST 2 COLUMNS. WHEN YOU PUNCH THE DECIMAL POINT, YOU OVERRIDE THE FORMAT FOR THAT FIELD. SOME VERSIONS OF THE INPUT ROUTINE FORBID BLANKS ON THE RIGHT OF A FIELD, SO ALL NUMBERS MUST BE RIGHT-JUSTIFIED IN THEIR FIELDS. HERE IS A MORE COMPLICATED FORMAT.

```
14 FORMAT('□',F14.1,3F5.0,3I5,E30.0,//,'OR THEREABOUTS')
```

THIS ONE IF USED FOR OUTPUT SAYS FIRST A BLANK CHARACTER, THEN A REAL NUMBER OCCUPYING 14 COLUMNS (INCLUDING BLANKS) AND HAVING 1 DIGIT TO THE RIGHT OF THE POINT, THEN THREE REAL NUMBERS EACH 5 COLUMNS WIDE AND WITH NO DIGITS TO THE RIGHT OF THEIR POINTS, THEN THREE INTEGERS EACH FIVE COLUMNS WIDE, THEN A REAL NUMBER IN E FORMAT WITH ITS EXPONENT SHOWING TAKING UP 30 COLUMNS IN ALL, THEN TAKE TWO LINES, AND ON THAT SECOND LINE A HOLLERITH FIELD SAYING 'OR THEREABOUTS.' EVEN MORE COMPLEXITY IS ALLOWED.

```
53 FORMAT(1H□, 3(2F10.5,6X))
```

SAYS FIRST A HOLLERITH FIELD CONSISTING OF ONE BLANK, THEN 3 GROUPS EACH HAVING 2 FIELDS OF THE KIND F10.5 AND A FIELD OF 6 BLANKS. THAT IS AS BAD AS THINGS CAN GET, BECAUSE IT IS NOT LEGAL TO COMBINE THE GROUPS INTO SUPER GROUPS. SEE ALSO E, F, H, I, X, SLASH, AND QUOTE.

SOME WRITE STATEMENTS WHICH MIGHT GO WITH THESE FORMATS ARE

```
WRITE(6,17) X,Y,(A(J),J=1,8)
WRITE(6,14) YOYO,AB1,AB2,AB3
WRITE(6,53) (Z(K),K=1,8)
```

ONLY THE FIRST OF THESE WRITE STATEMENTS FITS ITS FORMAT PERFECTLY, FOR THE SECOND USES LESS THAN THE FULL FORMAT AND THE THIRD USES MORE. THIS IS LEGAL. SEE READ STATEMENTS.

FUNCTION IS THE FIRST CARD OF A FUNCTION SUBPROGRAM. AN EXAMPLE IS

```
  FUNCTION SIGNM(X)
  IF(X)1,2,3
1 SIGNM=-1.
  RETURN
2 SIGNM=0.
  RETURN
3 SIGNM=1.
  RETURN
  END
```

OBSERVE THE STATEMENTS 1,2, AND 3 WHICH 'RETURN THE VALUE' OF THE FUNCTION. AFTER THIS PROGRAM IS PUT AWAY IN MEMORY, A MAIN PROGRAM MIGHT USE IT BY HAVING SUCH A CARD AS

```
U=V+5.*SIGNM(1.2+W)
```

JUST AS THOUGH SIGNM WERE IN THE LIBRARY. A FUNCTION MAY HAVE MORE THAN ONE ARGUMENT. HOWEVER, IT RETURNS ONLY ONE VALUE. SEE ALSO ARITHMETIC STATEMENT FUNCTIONS.

GO TO LOOKS LIKE

```
GO TO 88
```

OR ANY OTHER STATEMENT NUMBER. ONLY CONSTANTS MAY BE USED. THE CARD AFTER A GO TO MUST EITHER HAVE A STATEMENT NUMBER OR ELSE BE THE END CARD. SEE ALSO COMPUTED GO TO.

H FORMAT IS FOR A COUNTED HOLLERITH FIELD. FOR INSTANCE

```
17 FORMAT(13HTHE□ANSWER□IS,F20.8)
```

DESCRIBES A LINE BEGINNING WITH THE 13 CHARACTERS 'THE ANSWER IS' FOLLOWED BY A REAL NUMBER IN F FORMAT. HOLLERITH FIELDS ARE RARELY USED FOR INPUT, BUT THEY CAN BE SO USED. IN OUR EXAMPLE, THE FIRST 13 CHARACTERS OF THE INCOMING RECORD WOULD REPLACE OUR WORDING. OUR FORMAT COULD JUST AS WELL HAVE BEEN WRITTEN AS

```
17 FORMAT('THE□ANSWER□IS',F20.8)
```

HOLLERITH. SEE QUOTE.

I FORMAT IS FOR INTEGER VARIABLES. FOR EXAMPLE

```
17 FORMAT(6I7)
```

DESCRIBES A RECORD HAVING 6 INTEGER NUMBERS, EACH RIGHT-JUSTIFIED IN A FIELD 7 COLUMNS WIDE.

IABS SIGNIFIES THE INTEGER ABSOLUTE VALUE FUNCTION. IT TAKES AN INTEGER ARGUMENT AND DELIVERS AN INTEGER ANSWER.

IF STATEMENTS LOOK FOR EXAMPLE LIKE

```
IF(ABS(X-Y)-.01)1,2,2
```

OR MAYBE

```
IF(I-J)33,44,55
```

IF THE EXPRESSION IN THE PARENTHESES IS NEGATIVE GO TO THE LEFT STATEMENT NUMBER, AND IF THE VALUE IS ZERO GO TO THE CENTER NUMBER, AND IF THE VALUE IS POSITIVE GO TO THE STATEMENT NUMBER ON THE RIGHT. IF THE CARD FOLLOWING AN IF IS NOT AN END CARD AND HAS NO NUMBER, THE MACHINE WILL DECLARE AN ERROR.

IFIX IS A LIBRARY FUNCTION THAT WILL RETURN THE NEXT INTEGER LESS THAN OR EQUAL TO A NONNEGATIVE REAL ARGUMENT. IF THE ARGUMENT IS NEGATIVE, THEN IFIX RETURNS THE NEXT INTEGER THAT IS GREATER THAN OR EQUAL TO THE ARGUMENT. IN BOTH CASES IFIX TAKES A FLOATING-POINT ARGUMENT AND RETURNS THE INTEGER THAT IS CLOSEST TO THE ORIGIN. EXAMPLES ARE

```
        J=IFIX (Y)
WHERE   3=IFIX (3.984)
        0=IFIX (0.0)
       -4=IFIX (-4.83)
```

IMPLIED DO LOOPS ARE THINGS LIKE

```
WRITE(6,17) (X(J),J=2,NICKY)
```

THEY CAN NEST, TOO, AS FOR INSTANCE

```
WRITE(6,14) ((YAY(I,J),J=1, N, 3),I=IGO,8)
```

HERE SOME OF THE MEMBERS OF THE YAY ARRAY ARE BEING PRINTED. I AND J WILL NOT BE PRINTED BY THIS CARD. OF COURSE, IMPLIED DO LOOPS CAN BE USED WITH READING AS WELL AS WRITING.

INTEGER AND FIXED-POINT MEAN THE SAME THING. INTEGER NUMBER CONSTANTS NEVER HAVE DECIMAL POINTS. 1, 0, 234, -56, AND 32000 ARE EXAMPLES. THEY ARE SAID TO BE OF TYPE I. (THAT IS A LETTER, NOT A DIGIT.) INTEGER VALUES

INSIDE SOME MACHINES MAY NOT EXCEED ABOUT 32000 IN ABSOLUTE VALUE. THE INITIAL LETTERS I, J, K, L, M, AND N ARE RESERVED FOR INTEGER VARIABLES.

INTERNAL FUNCTIONS ARE ARITHMETIC STATEMENT FUNCTIONS, WHILE THE EXTERNAL FUNCTIONS AND SUBROUTINES HAVE PROGRAM DECKS OF THEIR OWN. OBVIOUSLY AN INTERNAL FUNCTION IS KNOWN ONLY TO ITS OWN DECK, BUT AN EXTERNAL FUNCTION OR SUBROUTINE IS KNOWN TO ALL THE OTHER DECKS LOADED WITH IT.

LIBRARY. FUNCTIONS ALREADY IN THE LIBRARY PART OF THE DISK OR TAPE INCLUDE SIN, COS, ATAN, EXP, ALOG, SQRT, ABS, AND IABS. ALL OF THESE EXCEPT THE LAST TAKE FLOATING ARGUMENTS AND DELIVER FLOATING VALUES. IABS TAKES A FIXED-POINT ARGUMENT AND DELIVERS A FIXED-POINT VALUE.

LOGARITHM. SEE ALOG.

LOOPS ARE FORMED BY GO TO'S, IF'S, COMPUTED GO TO'S, AND DO'S. THE DO LOOPS MUST NEST CORRECTLY.

MIXED MODE IS FOR INSTANCE X+J OR YAY*2 OR THE LIKE. SOME FORTRANS FORBID IT.

MODE. SEE TYPE.

NAMES OF VARIABLES, ARRAYS, FUNCTIONS BOTH INTERNAL AND EXTERNAL, AND SUBROUTINES ALL FOLLOW THE SAME RULES. A NAME CONSISTS OF SIX OR FEWER LETTERS AND DIGITS OF WHICH THE FIRST IS A LETTER. THE INITIALS I THROUGH N ARE RESERVED FOR THE FIRST LETTER IN THE INTEGER TYPE, AND THE OTHER LETTERS FOR THE FIRST LETTER IN THE REAL TYPE. OF COURSE, TYPE DOES NOT MEAN ANYTHING FOR SUBROUTINES.

NEST. IF TWO DO LOOPS SHARE ONE OR MORE CARDS, THEN ONE OF THE DO LOOPS MUST BE A PART OF THE OTHER, FOR EXAMPLE

```
  DO 1 J=1,3
  DO 2 K=1,10
2 X(J,K+1)=0.
1 CONTINUE
```

SUCH AN ARRANGEMENT IS CALLED A NEST OF DO LOOPS. ALL THE DO LOOPS IN A NEST HAVE DIFFERENT CONTROL VARIABLES.

ORDER. AFTER THE MONITOR RECORDS AND FORTRAN RECORDS COME CHANGE-OF-TYPE CARDS, DIMENSION CARDS, COMMON CARDS, EQUIVALENCE CARDS, ARITHMETIC STATEMENT FUNCTION DEFINITIONS, THEN EXECUTABLE CARDS AND FORMATS MIXED IN TOGETHER, AND FINALLY THE END CARD AND ANY MONITOR RECORDS TO FOLLOW. THE ORDER OF EXECUTION IS FROM THE TOP OF THE DECK DOWNWARD, UNLESS A TRANSFER MAKES THE MACHINE GO SOMEWHERE ELSE. PLAINLY IF THE CARDS ARE OUT OF ORDER THEN THE MACHINE WILL MAKE HASH OF THE ARITHMETIC.

PI=4.*ATAN(1.), IF YOU MEAN THE RATIO OF CIRCUMFERENCE TO DIAMETER. BE SURE TO DO THAT ARCTANGENT CALL AT THE START OF THE PROGRAM BEFORE ENTERING ANY LOOPS, BECAUSE IT IS SLOW. IT IS POSSIBLY BETTER TO USE THE CONSTANT 3.14159265....

PRECISION IS NORMALLY ABOUT SIX DECIMAL DIGITS BUT MAY BE EXTENDED TO 12 BY USING A CARD READING FOR EXAMPLE DOUBLE PRECISION (X,Y,Z(20,20)). THIS CARD FOLLOWS THE FORTRAN RECORD CARDS AT THE BEGINNING OF A DECK PROGRAM.

PROGRAM. A PROGRAM IS A DECK OF FORTRAN SOURCE CARDS ENDING WITH AN END CARD. IT CAN BE A MAIN PROGRAM, A FUNCTION, OR A SUBROUTINE. ONE OR MORE PROGRAM DECKS AND THEIR CONTROL CARDS MAKE A JOB. HERE IS A PROGRAM.

```
   X=2.0+2.0
   WRITE(6,17) X
17 FORMAT('□'F20.8)
   CALL EXIT
   END
```

QUOTE FORMAT IS FOR A HOLLERITH FIELD. A HOLLERITH FIELD IS A COLLECTION OF GENERAL CHARACTERS. FOR INSTANCE

```
17 FORMAT ('THE □ ANSWER □ IS', F20.8)
```

DESCRIBES A LINE BEGINNING WITH THE 13 CHARACTERS (COUNTING TWO BLANKS) THE ANSWER IS FOLLOWED BY A REAL NUMBER IN F FORMAT. HOLLERITH FIELDS ARE RARELY USED FOR INPUT, BUT THEY CAN BE SO USED. IN OUR EXAMPLE, THE FIRST 13 CHARACTERS OF THE INCOMING RECORD WOULD REPLACE OUR WORDING.

READ STATEMENTS HAVE THE APPEARANCE

```
READ(5,17) X,Y,(A(J),J=1,10)
```

OR THE LIKE. HERE 5 MEANS THE CARD READER, 17 IS A FORMAT STATEMENT NUMBER AND THE VARIABLES X,Y,A(1), A(2),...,A(10) ARE TO BE READ IN. IF THE FORMAT CARD IS

```
17 FORMAT(F10.0,F5.0,10F1.0)
```

THEN X WILL GET THE FORMAT F10.0 AND Y WILL GET F5.0 AND EACH A(J) WILL GET AN F1.0 FORMAT. IT IS NOT NECESSARY, BY THE WAY, FOR THE READ STATEMENT AND THE FORMAT TO FIT EXACTLY. IF THE FORMAT IS TOO SMALL, IT WILL START TO REPEAT ON A NEW CARD, AND IF IT IS TOO LARGE THEN THE LATTER PART WILL JUST NOT GET USED. IF THE DIMENSION CARD FOR A (J) MADE IT 10 LONG, THEN WE COULD WITH THE SAME EFFECT WRITE

```
READ(5,17) X,Y,A
```

AS AN ABBREVIATION. THE SAME TRICK MAY BE TRIED WITH MATRICES, BUT THEY COME OUT TRANSPOSED. USE NESTED DO LOOPS OR IMPLIED DO LOOPS TO READ IN A MATRIX. FOR EXAMPLE

```
      READ (5,27) ((Y(I,J), J=1,5), I=1,8)
   27 FORMAT (5F10.4)
```

WILL READ IN AN 8×5 MATRIX ONE ROW OF 5 NUMBERS AT A TIME.

REAL AND FLOATING-POINT ARE THE SAME THING. REAL NUMBER CONSTANTS ALWAYS HAVE A DECIMAL POINT. EXAMPLES IN F FORM ARE 4., 4.5, .056, 789456.00, AND 0., WHILE EXAMPLES IN E FORM ARE 1.E4, 2.3E-11, AND 12345.E12, WHERE WE UNDERSTAND THE E TO PRECEDE A POWER TO WHICH 10 IS RAISED. THE LAST THREE NUMBERS ARE HENCE 1. TIMES 10 TO THE 4, 2.3 TIMES 10 TO THE MINUS 11, AND 12345. TIMES 10 TO THE TWELVE. BOTH KINDS OF CONSTANTS MAY BE USED IN EXPRESSIONS. REAL VARIABLES BEGIN WITH LETTERS BEFORE I OR AFTER N.

REPLACEMENT STATEMENT. THIS IS THE MOST COMMON CARD IN FORTRAN. IT REPLACES THE NUMBER ON THE LEFT-SIDE OF AN EQUALITY WITH THE COMPUTED VALUE OF THE EXPRESSION ON THE RIGHT-SIDE. SOME EXAMPLES ARE

```
X=1.+X+Y
DIST=RATE*TIME
U(I,J)=XA(I)*XB(J)-SQRT(H)
```

THE LEFT-HAND SIDE MAY BE A VARIABLE OR SUBSCRIPTED VARIABLE. THE RIGHT-HAND SIDE MAY BE ANY EXPRESSION. THE TWO SIDES MAY DISAGREE IN TYPE, AS

```
YNEW=J+IABS(MUMU-KAW)
```

OR

N=X+.5

IN CHANGING FROM FLOATING TO FIXED-POINT, THE MACHINE WILL LOP OFF THE FRACTION, SO IF X WAS 2.9 THEN N WILL BECOME 3 WITH NO FRACTIONAL PART. IF HOWEVER X WAS −2.9 THEN N WILL BECOME −2 WITH NO FRACTIONAL PART. NOTICE THAT

X=X+1.

MAKES GOOD SENSE IN FORTRAN. IT MEANS ADD 1. TO THE OLD VALUE OF X TO GET THE NEW VALUE.

KEEP IN MIND THAT FIXED-POINT NUMBERS LARGER THAN ABOUT 32000 IN ABSOLUTE VALUE CANNOT EXIST INSIDE SOME MACHINES, SO

N=1000000

MAY GIVE A WRONG ANSWER. CHECK ON YOUR MACHINE.

RETURN LOOKS LIKE

RETURN

AND IS FOUND ONLY IN SUBROUTINE AND FUNCTION DECKS. IT MEANS JUMP BACK TO THE PROGRAM THAT CALLED THIS PROGRAM. HERE IS AN EXAMPLE.

```
      SUBROUTINE CHECK(DIVSR)
      IF(DIVSR)1,2,1
    1 RETURN
    2 WRITE(6,17)
   17 FORMAT('□', 'HERE COMES ANOTHER ZERO, FOLKS.')
      RETURN
      END
```

SIGN MEANS A PLUS OR A MINUS. A NUMBER HAVING A SIGN IS CALLED SIGNED. IF A NUMBER IS UNSIGNED, WE UNDERSTAND IT AS THOUGH IT HAD A PLUS SIGN.

SIN SIGNIFIES THE TRIGONOMETRIC SINE FUNCTION. THE ARGUMENT MUST BE IN RADIANS.

SLASH. FOR EXAMPLE

```
17 FORMAT(/////,'□',5F20.8)
```

USED FOR PRINTER OUTPUT SAYS TAKE FIVE LINES AND THEN MAKE A BLANK FOLLOWED BY 5 REAL NUMBERS IN F FORMAT. ON INPUT, SAY THE CARD READER, THE SLASH MEANS TAKE A NEW CARD.

SOFTWARE MEANS THE PROGRAMS SENT FROM THE MACHINE DESIGNERS OR THOSE SUPPLIED BY OUR SYSTEM PROGRAMMERS. IT INCLUDES THE COMPILER, THE ASSEMBLER,

THE LOADER, THE LIBRARY, THE MONITOR CONTROL RECORD ANALYZER, AND THE SKELETON SUPERVISOR.

SOURCE PROGRAMS ARE WRITTEN IN FORTRAN OR SOME OTHER LANGUAGE. ACTUALLY THE MACHINE TRANSLATES THIS INTO ANOTHER FORM CALLED THE OBJECT PROGRAM.

SPACE. THE PRINTER IS CAUSED TO SPACE 1 LINE IN THE USUAL WAY BY A BLANK CONTROL CHARACTER. A ZERO MAKES A DOUBLE-SPACE, AND A PLUS SIGN STRIKES THE SAME LINE WITHOUT SPACING UP. A BLANK CHARACTER IN A FORMAT MAY BE WRITTEN BY '□' OR 1H□ OR 1X. FIVE BLANKS CAN BE WRITTEN BY 5X OR '□□□□□' OR 5H □□□□□.

STATEMENT NUMBERS GO IN COLUMNS 1 THROUGH 5. THEY ARE UNSIGNED FIXED-POINT CONSTANTS OF 5 OR FEWER DIGITS. ONLY EXECUTABLE STATEMENTS AND FORMATS NEED NUMBERS. OTHER CARDS WILL HAVE THEIR STATEMENT NUMBERS STRIPPED OFF. IF A CARD IS NOT A FORMAT OR THE TARGET OF A TRANSFER, IT DOES NOT NEED A NUMBER. THE NUMBERS NEED NOT BE IN ANY SPECIAL ORDER.

STATEMENTS USUALLY GO ONE TO A CARD. A LONG STATEMENT MAY NEED MORE THAN ONE CARD. IN THAT CASE, STOP AT COLUMN 72, PUNCH A NONZERO NONBLANK CHARACTER IN COLUMN 6 OF THE NEXT CARD, AND PROCEED. IN THIS WAY A STATEMENT MAY OCCUPY 5 CARDS INCLUDING THE FIRST. IT IS NOT LEGAL TO CONTINUE A COMMENT STATEMENT IN THIS WAY.

STOP. EXECUTION OF A STOP CARD KILLS THE PROGRAM. EXAMPLE

```
      STOP
```

ANOTHER WAY TO KILL THE PROGRAM IS TO USE THE CALL EXIT CARD.

```
      CALL EXIT
```

SUBROUTINE IS THE FIRST NONCOMMENT CARD OF A SUBROUTINE PROGRAM. FOR EXAMPLE

```
      SUBROUTINE VECAD(A,B,C)
      DIMENSION A(10),B(10),C(10)
      DO 56 J=1,10
   56 A(J)=B(J)+C(J)
      RETURN
      END
```

IS A SIMPLE VECTOR ADDITION PROGRAM. A MAIN PROGRAM TO CALL IT MIGHT LOOK LIKE

```
      DIMENSION X(10),Y(10),Z(10)
      READ(5,17) Y,Z
   17 FORMAT(10F8.2)
      CALL VECAD(X,Y,Z)
      WRITE(6,18) X
   18 FORMAT('□',10F12.6)
      CALL EXIT
      END
```

SUBSCRIPT. A VARIABLE HAS THE SAME NUMBER OF SUBSCRIPTS ON OTHER CARDS THAT IT HAD ON THE DIMENSION CARD. FOR EXAMPLE

```
DIMENSION X(2,56),YACHT(8)
```

MAY BE FOLLOWED LATER IN THE PROGRAM BY

```
X(I+1,J)=1.2+4.*YACHT(J-8)
```

X IS SEEN TO HAVE 2 SUBSCRIPTS AND YACHT HAS 1. THE VALUE OF J−8 HAD BETTER BE A NUMBER FROM 1 TO 8 INCLUSIVE, OR CHAOS WILL RESULT. SIMILARLY I+1 MUST BE 1 OR 2, AND J MUST LIE BETWEEN 1 AND 56 INCLUSIVE. ONLY A FEW SIMPLE KINDS OF EXPRESSIONS MAY BE USED AS SUBSCRIPTS. THEY ARE OF THE KINDS

```
J
J+17
J-17
3*J
3*J+17
```

AND

```
3*J-17
```

WHERE ANY NONSUBSCRIPTED INTEGER VARIABLE MAY BE USED INSTEAD OF J AND ANY UNSIGNED INTEGER CONSTANTS MAY BE USED FOR 3 AND 17. ALL OTHER FORMS ARE STRICTLY FORBIDDEN. ONE MAY NOT, FOR EXAMPLE, TRY TO USE J*3 AS A SUBSCRIPT, FOR IT OUGHT TO HAVE BEEN 3*J. ALSO 1+J IS WRONG FOR J+1 AND SO ON. NOTHING AT ALL CAN BE DONE IN THE WAY OF I+J, EXCEPT ON TWO CARDS AS

```
K=I+J
YACHT(K)=4.5
```

SYNTAX. THE STRUCTURE OF EXPRESSIONS IN A LANGUAGE SUCH AS FORTRAN OR ENGLISH IS CALLED SYNTAX. SYNTAX ALSO REFERS TO THE RULES GOVERNING THE STRUCTURE OF A LANGUAGE.

TRANSFER. THE TRANSFER STATEMENTS OF FORTRAN INCLUDE GO TO, IF, COMPUTED GO TO, AND DO. IT IS FORBIDDEN FOR A DO LOOP TO END ON A TRANSFER STATEMENT OR A FORMAT.

TYPE. THE TWO TYPES ARE INTEGER AND REAL. SEE INTEGER AND SEE REAL. IT IS POSSIBLE TO CHANGE THE TYPE OF A VARIABLE OR ARRAY. FOR EXAMPLE

```
INTEGER X, Y(3)
REAL J, K(2,2)
```

WILL CHANGE X TO INTEGER AND Y TO INTEGER AND DIMENSION Y, AND SO ON. IF A VARIABLE IS TO BE BOTH TYPE-CHANGED AND DIMENSIONED, THEN THE DIMENSIONING MUST BE DONE ON THE CHANGE-OF-TYPE CARD AS SHOWN HERE, AND NOT ON A DIMENSION CARD. ALL CHANGE-OF-TYPE CARDS SHOULD PRECEDE ALL DIMENSION CARDS.

UNDEFINED VARIABLE. FOR INSTANCE, IF THE BEGINNING OF A PROGRAM IS

```
X=Y
```

THEN HOW COULD THE COMPUTER POSSIBLY KNOW WHAT TO USE FOR Y. Y IS AN UNDEFINED VARIABLE. SOMETIMES THIS KIND OF ERROR GETS BY THE MACHINE. FOR INSTANCE IF THE VARIABLE X FIRST APPEARS AS

```
X=X-(X*X-2.)/(2.*X)
```

THEN THE MACHINE MAY TRY TO EXECUTE THE PROGRAM ANYWAY, BUT WRONG ANSWERS ARE LIKELY. IF A DUMMY VARIABLE OF A FUNCTION OR SUBROUTINE IS NOT USED THEN THE COMPILER WILL SAY IT IS UNDEFINED, BUT PERHAPS, IN THIS CASE, THE SYSTEM IS TAKING TOO MUCH ON ITSELF.

UNREFERENCED STATEMENT. YOU PUT A STATEMENT NUMBER ON A CARD AND NEVER REFERRED TO THAT NUMBER ELSEWHERE IN THE PROGRAM. THIS IS NOT AN ERROR, BUT THE COMPILER IS SUSPICIOUS AND REPORTS IT FOR YOUR INFORMATION.

VARIABLES HAVE NAMES OF SIX OR FEWER LETTERS OR DIGITS STARTING WITH A LETTER. INITIALS I THROUGH N ARE FOR THE FIRST LETTER OF FIXED-POINT VARIABLES, AND THE OTHERS FOR THE FIRST LETTER OF FLOATING-POINT VARIABLES. HENCE, B, YACHT, AND W123S ARE FLOATING-POINT VARIABLES, AND J, KAW, AND M7756 ARE FIXED-POINT VARIABLES.

WRITE STATEMENTS WORK LIKE READ STATEMENTS EXCEPT THAT THE DATA COME OUT OF THE MACHINE INSTEAD OF GOING IN. A TYPICAL WRITE IS

```
WRITE(6,18) X,Y,(A(J),J=1,10)
```

THE 6 STANDS FOR THE LINE PRINTER. YOUR MACHINE MAY USE SOME CHANNEL OTHER THAN 6.

FOR MORE DETAILS SEE THE REMARKS AT READ.

X FORMAT IS FOR MAKING BLANKS. FOR INSTANCE

```
17 FORMAT(23X,F10.5)
```

DESCRIBES A LINE HAVING 23 BLANKS FOLLOWED BY A REAL NUMBER IN F FORMAT. ON INPUT THE FIRST 23 COLUMNS WOULD BE SKIPPED OVER.

INDEX

1 2 3 4 5 6 7 8 9 10 11 12 13 14 15 16 17 18 19 20 21 22 23 24 25   80 79 78 77 76 75 74 73 72